Scientific Writing

Thinking in Words

=

David Lindsay

CSIRO
PUBLISHING

D0068159

National Library of Australia Cataloguing-in-Publication entry

Lindsay, D. R.

    Scientific writing = thinking in words / by David Lindsay.

    9780643100466 (pbk.)
    9780643101579 (ePdf)
    9780643102231 (ePub)

    Includes index.

    Technical writing – Study and teaching.

    Communication of technical information.

808.0665

Published by

CSIRO PUBLISHING
150 Oxford Street (PO Box 1139)
Collingwood VIC 3066
Australia

Telephone: +61 3 9662 7666
Local call: 1300 788 000 (Australia only)
Fax: +61 3 9662 7555
Email: publishing.sales@csiro.au
Web site: www.publish.csiro.au

Front cover image concept by Kate Lindsay.
The photographs on pages 93 and 94 are by iStockphoto.

Set in 9/12 Palatino
Cover design by Modern Art Production Group
Printed in China by 1010 Printing International Ltd
Reprinted in Australia by Ligare, 2011, 2012

CSIRO PUBLISHING publishes and distributes scientific, technical and health science books, magazines and journals from Australia to a worldwide audience and conducts these activities autonomously from the research activities of the Commonwealth Scientific and Industrial Research Organisation (CSIRO).

Original print edition:
The paper this book is printed on is in accordance with the rules of the Forest Stewardship Council®. The FSC® promotes environmentally responsible, socially beneficial and economically viable management of the world's forests.

Contents

Preface V

Thinking about your writing 1

Getting into the mood for writing 3

What is a 'good' style for scientific writing? 4

The fundamentals of building the scientific article............................. 6

Getting started 9

Writing about your thinking...15

The *Title*..17

The *Introduction* 20

The reasoning behind the hypothesis—the other part of the *Introduction*..................25

The *Materials and Methods* 28

The *Results* 31

What to present...32

What form of presentation? Tables, figures or text? 34

Graphs or tables? 36

Use of statistics in presentation of results 38

The *Discussion* 39

What makes an effective *Discussion*? 39

What is there to discuss? 41

Giving impact to your scientific story 42

The paragraph as a vehicle for your arguments 44

Speculation in the *Discussion*...47

The length of the *Discussion*...47

Citations in the *Discussion*...48

Checking the logic of the *Discussion*...49

The *Summary* or *Abstract*...49

Constructing the *Summary* 50

The other bits...51

Authorship 51

Acknowledgements 53

The *Bibliography*..53

Editing for readability and style 55
    Eliminating verbal stumbling blocks 56
    The seven verbal stumbling blocks... 56
    Delivering the written word in a way that matches the way a reader reads 64
    Where to from here?... 68
    Final editing for style... 69
    Choosing the journal.. 71
    Sending to the journal.. 72
    Coping with editors, referees and reviewers 72
    Re-submitting to the journal... 74

Thinking and writing beyond the scientific article 77
    The text for oral presentation at a scientific seminar................................ 78
        Structure... 78
    Design and preparation of posters for conferences 88
        What makes a successful poster?... 89
        The structure of a successful poster 90
    The review 95
        The structure of the review 96
        New ideas.. 97
        The literature 98
        Being specific... 98
        Some common difficulties with reviews... 99
    Writing science for non-scientists... 100
        What a reader wants to read and a scientist wants to say......................... 101
        What makes a good article? 102
        The essential ingredients 104
        Constructing the article 105
        The final inspection 106
    The thesis... 106
        Form and layout of a thesis 107
        Review of the literature in the thesis 107
        Getting down to business in writing the thesis—the working summary.............. 115
        Using the working summary 116

    Index 118

Preface

HUNDREDS OF PEOPLE contributed to this book. Most of them were researchers who attended workshops and courses in which we collectively applied concepts about thinking and reasoning to the task of converting ideas and experimental data into focused articles for publication. They came from many countries and spoke many languages. They tested the concepts to the limit in subjects that ranged from complex molecular biology to marketing and legal practice and almost everything else in between. From this emerged the principles of thinking and writing that the book illustrates and I am grateful for their robust challenges and views because I cannot recall one workshop in which I did not learn something new or modify something that I thought was indisputable.

Scientific writing is dynamic. For proof, you only have to compare a modern-day article with one written, say, in the 1960s. Of course, some things such as the need for precision, clarity and brevity seem to be immutable, but many others, like the use of the passive voice or the first person—I or we—have changed remarkably in a relatively short time. The electronic era has altered and will continue to alter the way articles are submitted, reviewed and even read. But the necessity for good writing is as strong as ever. However, to keep up with these changes, I will need to revise this book periodically and I need your help. Somewhere in this book I use the cliché that the perfect scientific article is yet to be written. That applies equally to books, but an inherent catch in writing a book about writing is that it primes the reader to recognise its faults more easily than a book on other subjects. So, you, the reader, are better placed than most to advise on how to improve this book and I welcome your comments should you be moved to make them.

Then there are my colleagues who use the principles of structure and style regularly in their own work and teaching but never hesitate to open vigorous discussions in improbable locations and at extraordinary times to question some aspect or another. Foremost among them are Pascal Poindron, a Frenchman fluent in English and Spanish, Pierre Le Neindre, another Frenchman fluent in English, and Ian Williams, an Australian colleague, passionate about good writing, who all made valuable additions and modifications to the many drafts. In addition, they made me acutely aware of the problems, and sometimes advantages, that arise when authors who do not have English as their native tongue are compelled to write their work in English which, by chance, happens to be the de facto, universal language of science. As a result, it compelled me to address many of the aspects of scientific writing from the viewpoint of non-native English speaking authors and to emphasise that they are not as disadvantaged as they perhaps may think. The language of science which conveys logic and reasoning, is independent of the language in which it happens to be expressed. Since the primary goal of good scientific writing is to communicate good science, non-native English speakers who are good scientists have all the tools they need to write well although they may need some help eventually to tidy it up for publication in English-language journals.

I am indebted to my daughter, Kate, for her professional layout of the material in the book and for the concept of the design of the cover and to my wife, Rosalind, for countless times she mostly willingly perused and corrected the drafts.

David Lindsay (September 2010)

Thinking about your writing

TELLING PEOPLE ABOUT RESEARCH IS JUST AS IMPORTANT AS doing it. But many researchers, who, in all other respects, are competent scientists, are afraid of writing. They are wary of the unwritten rules, the unspoken dogma and the inexplicably complex style, all of which seem to pervade conventional thinking about scientific writing. In this section, we bring these phantoms into the open, expose them as largely smoke and mirrors, and replace them with principles that make communicating research easier and encourage researchers to write confidently.

Getting into the mood for writing..3

What is a 'good' style for scientific writing? 4

The fundamentals of building the scientific article 6

Getting started..9

ONE OF THE GREATEST PARADOXES IN RESEARCH IS THAT, regardless of the field, work must be written and published before it can be considered complete, yet training in writing is rare in the training curriculum of budding scientists.

There is a common saying, 'If you haven't written it, you haven't done it.' A research project is not complete just because the last sample has been taken or the last set of data analysed. If you are in the world of research, it is of little value to have a colleague or two in the next office or laboratory know that you have discovered something. From the day that you completed undergraduate training and decided to become a researcher, your circle of colleagues or potential colleagues expanded from being a relatively few fellow students to an indefinite number of fellow researchers from all over the world. Communicating with them is a very different task from the one you were involved in as a student. In fact, you may have to spend as much time writing, reading or correcting manuscripts as you do on research itself. Even if you have told delegates at a large meeting or a convention what you have done, the proportion of scientists in your field that were there and listening to what you said is tiny and probably transient. 'The spoken word evaporates but the written word stays on.' The written word is permanent, all pervading and the best way to tell the world of research that you are a noteworthy part of it.

If you haven't written it, you haven't done it …

Despite this, writing is one of the most inadequately developed of all the skills that scientists use in their research activities. Let us look briefly at the statistics.

- 99% of scientists agree that writing is an integral part of their job as scientists
- Fewer than 5% have ever had any formal instruction in scientific writing as part of their scientific training
- For most, the only learning experience they have is the example they get from the scientific literature that they read
- About 10% enjoy writing; the other 90% consider it a necessary chore.

These figures are, of course, approximate but they come from informal surveys conducted over many years in many countries and, I believe, are close to reality.

Beneath these statistics, it is easy to deduce a serious problem. For example, if 90% of scientists do not really enjoy writing then most of the scientific literature in front of us is written by people who did not enjoy writing it. The chances are that, regardless of the quality of the science, it has been cobbled together to get it published, reviewed by referees who have little more interest or knowledge about writing clearly than the authors and, finally, published in a style that has had little critical review. Thus, a big proportion of the literature on which developing scientists base their ideas of writing style and structure has been written and reviewed by people who knew little about style and structure and probably didn't enjoy writing anyway. That is not an effective model because it is highly variable and, on average, not very good.

We can develop the common saying further. 'If you write it, but no one reads it, you still haven't done it.' The only reason for writing is to have what you have written read and understood by other scientists and this is often forgotten by scientists when they commit their work to paper. They believe, and are often encouraged to believe, that publication in a journal is the ultimate end-point for a piece of research. It is not. The paper must then be read and understood clearly by the scientific community around the world in the relevant and related fields before the job can be deemed to have been completed successfully. So, we can extend the saying even further, 'If you write it up and it is read but not understood you still haven't done it.'

By contrast to the many bad models of writing that we come across, there are some beautifully written and structured papers that stand out like beacons because they are so clear to read and deliver their scientific message so forcefully. These are the models that we must attempt to follow. Unfortunately, they also stand out because they are so rare.

If you write it, but no one reads it, you still haven't done it.

The suggestions for better writing in this book draw directly and indirectly on these outstanding models and are usually presented as principles rather than rules. It is up to you to decide if the principles make sense to you and, if they do, you can follow the further suggestions to modify the structure or the style of your writing to ensure that you are adhering to those principles.

Getting into the mood for writing

There seem to be two contrasting attitudes to the writing and discussion of scientific results. One is the positive attitude: 'I have just been part of an adventure of discovery in science and I have found something that I want to share with you, the reader. In this article, I am going to take you on the same adventure and tell you what made me excited about it. In doing so I hope you will recognise and appreciate my scientific contribution.'

The other is far more passive and, regrettably, seems to be more common. 'Research is the seeking and discovery of information that was not known previously. I am writing this for you, who have been trained a scientist to seek out information and make something of it. I am putting the data before you, together with some interpretation and I expect you to use your skills to work out much of what it means.' This description of the second approach may seem harsh but I believe that it is a fair interpretation of the way that many modern scientific articles are presented to readers.

If scientific articles are written to be read then it is important for you as a writer to have a realistic impression of the sort of person who is likely to be a reader and how they go about reading. In reality, potential readers are not likely to be motivated much differently from you. That means that they are busy, they have other things than reading scientific articles on their daily agenda and they will be happy to convince themselves that they don't need to read many of the articles in the journals that cross their desk. They certainly will not be reading articles just in case they contain some unforeseen but useful material hidden in some obscure paragraph. First, you have to attract their attention and then try to hold that attention until the last full stop. That should be your goal but, even with a well-written article, it is unlikely that you will often achieve it. At least in the first instance, readers are selective until they get a feeling for the article and what it has to offer them. Then, if it really interests them, they will come back and scrutinise the whole article carefully and with scientific interest. The challenge is to make sure that even if they spend just a few moments perusing your article, they will pick up the essentials of what it has to say. This means that they must find the most important parts clearly presented and in the places where they expect to find them. If they are forced to find your most interesting data buried in a heterogeneous mass of information in the *Results* or your most brilliant inspiration among a series of problematical comments in the *Discussion*, you will have little chance of having your work acknowledged or appreciated.

To write a paper succesfully, you have to do more than commit your data and comments to paper; you must work hard to ensure that your data and comments are structured and presented so that the reader has easy access to them.

What is a 'good' style for scientific writing?

In writing scientific articles, many of us struggle to achieve a style of writing that does not come naturally to us. We imagine that we must follow a convoluted style based on vague impressions of what we read in the scientific literature. Nothing could be further from the truth and it is here that many of the models that we use in the literature let us down.

There are just three immutable characteristics of good scientific writing that distinguish it from all other literature. It must always be

- precise
- clear
- brief

... and in that order. If it is vague, it is not scientific writing; if it is unclear or ambiguous, it is not scientific writing and if it is long winded and unnecessarily discursive, it is poor scientific writing. But do not sacrifice precision or clarity in order to be brief. So, if it takes a few more words to make what you want to say crystal clear to as many readers as possible, then use those words.

The good news is that, if you are precise, clear and brief, then you do not have to conform to any other specific rules to be a good scientific writer. The style of scientific writing is plain and simple English similar to that you would use in a conversation with a colleague. Or, as one author put it, 'The best style is no style at all.' That is also good news because it is the style with which we are most familiar and most skilled. We use it every day, we get constant feedback on how successful we have been in getting across what we want to say and we are therefore confident with it. When writing about research, we often have to explain procedures and concepts that are complex. So, it makes sense not to add further complexity by struggling with words and expressions that are unfamiliar to both the writer and the reader in order to conform to some imaginary style. Of course, you may decide that you want to impress your readers with your knowledge and command of English. If so, think again. You should be writing to inform, not impress. Sometimes, when I say this to young scientists, they ask whether editors or reviewers might think them naïve and unscientific if they use simple language. I can't speak for all editors and reviewers, but I cannot imagine any of them complaining that authors were not obscure enough in explaining themselves. If you are a scientist and your ambition is to gain the Nobel Prize one day, then try to get it for your science and not for your literature.

If you are a scientist and your ambition is to gain the Nobel Prize one day, then try to get it for your science and not for your literature.

There is another reason for writing in plain, simple English rather than using flowery, ornate or obscure prose. The language in which modern science is written is English yet, depending on the field, up to 50% of the scientists who may read a scientific article may not have English as their first, spoken language. If these people are discouraged by having to search for their dictionaries to understand what native English speakers have to say, the whole purpose of writing the article—to have it read and understood—will be totally lost. In fact, with the increasing spread of scientific expertise around the world, native English speakers have a serious obligation to their non-English speaking colleagues not to flaunt their good luck by inserting obscure words and expressions. Such words and expressions may be impressive, but for the wrong reasons.

Remember, your primary aim when writing a scientific article should be to have as many people as possible read it, understand it and be influenced by it.

The fundamentals of building the scientific article

Most people associated with science and research agree that writing and publishing the article that describes their experiment is an integral part of the research process. Unfortunately, many think that this process is accomplished in three distinct phases; planning the work, doing the research and writing it up. That is a pity, because all three phases are so closely integrated that none can be completed successfully without involving the other two.

The relationship between good planning and the smooth execution of a research program is obvious but the importance of thinking about writing the article during the planning is often overlooked. The title of this book *Scientific writing = thinking in words* came from the conviction that thinking and reasoning at the planning stage facilitate both the experimental and writing stages and, if well thought through, writing an experiment can be as stimulating as doing the experiment itself and certainly not, as many people seem to feel, a necessary but unpleasant task.

... thinking and reasoning at the planning stage facilitate both the experimental and writing stages.

Broadly, the thinking process in writing a paper parallels that for designing the experiment itself. It can be summarised like this:

Step 1. You predict the results of the research you are planning to do.

Step 2. You sort out why you think that you will get these results.

Step 3. You imagine how you would present them.

Step 4. You imagine how you would explain them.

At first, this may seem to be quite simple but in reality the thinking necessary to come up with satisfactory answers at each of these four steps is, probably, about three-quarters of all of the thinking that you will do for the whole writing process. And, doing this thinking before you start the experiment, and not when you have the results in front of you, ensures that you have the best chance of having convincing data with which to work. It reduces the risk of having to reproach yourself for poor planning; not having had another treatment group, or not having composed a supplementary question in the survey, or not measuring another factor in the analysis, any of which may have made the presentation of the results or the drawing of the conclusions more straightforward and more credible. It reduces the frustration of having a story to tell but having it compromised by the need to explain why your data were less convincing than they could have been.

However, it takes time and effort to work your way through these four steps. It may seem as if the prediction in Step 1 is simply a piece of guesswork but Step 2 quickly dispenses with that idea because it requires that you support your prediction with a logically reasoned case based on defensible evidence from published and acceptable information. This part, obviously, involves you in a lot of thinking, reading, interpretation and rethinking—and it takes time. The reward for this is that, once you have substantiated your prediction in this way, it becomes your hypothesis and you now have it as the central focus for the experiment you will do and write about.

So, there are a lot of advantages in the four-step process. You are compelled to think before you act, which is always a good thing. More important, you are compelled to think scientifically and logically before you act, which means that you are likely to be doing and writing about good science. Your writing will have a clear focus and that focus will lead to readers predicting what they are about to read which, in turn, makes reading an easy task.

There are many texts on the philosophy of science and scientific method that deal extensively with the hypothesis but, in short, we can describe it as 'a reasonable scientific proposal'. It is not a statement of fact but a statement that takes us just beyond known information and anticipates the next logical step in a sequence of supportable precepts. The hypothesis has to have two attributes to be useful in scientific investigation: it must fit the known information and it must be testable. To comply with the first attribute, you, the scientist have to read and understand the literature. To comply with the second, you have to do an experiment. In essence, the paper you are about to write concerns nothing other than those two things. You can see why the hypothesis is so central to scientific writing.

The supporters of the so called 'scientific method' tell us that the formulation, justification and testing of an hypothesis is basic to all worthwhile scientific research. What is less appreciated, but vital to what this book is all about, is that the hypothesis also has a key role in the written paper and is an essential ingredient in your thinking and your writing of that paper. This is because

- You have to know all of the known and acceptable information before you can propose an hypothesis.

- You save time and money by making many of your mistakes mentally before you commit yourself to doing the research.

- It gives your research a clear focus and, when you write up the research you, too, will have a clear focus.

Expressing your hypothesis in the *Introduction* is the most effective way of establishing that focus because it gives your readers a clear idea of what to expect in the rest of the scientific article. From the readers' point of view, this makes reading easier and much more pleasurable. From your point of view, as the author, it means that the reader will be able to follow your results and the arguments in your discussion from the same viewpoint as you.

Once you have put into words a well-reasoned hypothesis, the main part of the scientific article is, in fact, disarmingly simple to structure logically and with confidence. This is best explained in broad terms by looking at how the hypothesis influences the structure of three of the most important parts of your article, the *Introduction*, the *Results* and the *Discussion*.

The *Introduction* consists of just two parts: 1. the hypothesis or what you expected to find and 2. the logical reasoning that made this hypothesis the most plausible expectation about the phenomenon you were studying—and practically nothing else. Occasionally, these two essential elements may be backed by one or two sentences that put the work in context or emphasise its importance.

The *Results* can be given priorities rather than appear as a homogenous array of information. Results with high priorities are those that relate to the testing of the hypothesis and those of low priority are those that do not. When presented with these priorities in mind the results become immediately more meaningful and relevant to both writer and reader.

Once you have put into words a well-reasoned hypothesis, the main part of the scientific article is, in fact, disarmingly simple.

The *Discussion* can be organised similarly into components (or arguments) of different priorities based on whether or not they are about results that support or reject the hypothesis.

Let us assume that we are writing an article in which we propose an hypothesis and finish up accepting it. The article would take the following form:

The *Introduction* explained why this hypothesis was the most plausible expectation about the subject being explored

The *Results* backed this up

The *Discussion* explored the consequences in relation to the work of other researchers and, possibly, for broader application, either practical or theoretical.

The outcome is a good, well integrated and focused article.

But, good experiments are not just those that give rise to your accepting the hypothesis. What about the structure of an article in which the proposed hypothesis turned out to be wrong when it was tested?

The *Introduction* explained why this hypothesis was the most plausible expectation about the subject being explored … before you came up with these new results.

The *Results* blew a hole in that plausibility.

The *Discussion* explored why the logic that made the original hypothesis seem plausible was wrong, how we have to rethink our concepts about the work of others and, possibly, what we should do differently in applying this information practically or in theory.

In other words, a disproved hypothesis results in an equally good or even better paper than one supporting an hypothesis. Experiments designed around the development and testing of an hypothesis yield scientifically rewarding information regardless of whether the actual results match the expected ones. Writing the article follows the same path—you tell the reader what you expected to find, and why. Then, you present your findings and discuss how your findings matched your expectation.

So, in a nutshell, writing a good scientific article is as much an exercise in clear and focused thinking as it is in clear and accurate writing. But to be really successful, you have to have readers thinking along the same path as you when they read. And, to do that, you have to plan the structure of your article carefully.

Getting started

… your problem is not so much how you are going to start, but how you are going to finish.

Most scientists have a problem getting their writing process under way. To them, there is little more daunting than a blank screen on a computer or a clean sheet of paper waiting to be filled coherently and legibly with information and wisdom. They gather around them laboratory and field notebooks, printouts of statistical procedures, excerpts from papers by other scientists and notes and ideas scribbled on bits of paper and hope for an inspirational opening. During this gathering process, it is often a relief to be distracted by the telephone's ringing or a colleague calling in. These distractions may temporarily relieve the anxiety but they don't resolve the problem. It is handy to have a few, more effective strategies to get you under way.

The first step in getting started is to realise that your problem is not so much how you are going to start, but how you are going to finish. You would never knowingly set out on a major voyage without knowing your destination. Yet so often, when we set out on a voyage of writing, we jot down a few words and hope to be inspired somehow about the right direction to follow with all the words that ensue. The chances of that happening are very, very low. On the other hand, to know with some certainty, at the outset, how to finish an assignment as complex and demanding as a complete scientific paper is asking a lot more than most of us can manage. The secret is to reduce the scale of the task by breaking it into manageable sections. Then, by concentrating on these sections individually, deciding on appropriate conclusions to each of them and filling in the words that lead to that conclusion, you can progress efficiently. Later, as you understand more clearly where the whole article is headed, the sections can be amalgamated and edited to become a united whole that is consistent and coherent.

Fortunately, scientific papers have a relatively rigid physical structure that must be followed and this structure provides a primary breakdown into smaller sections. The IMRAD format, or *Introduction, Materials, Results* and *Discussion*, makes available four major elements, each with a different purpose and content that can be planned and written independently from the others—at least in the first instance. These, too, can be broken temporarily into components to let you, the writer, develop a mental image of what you want to say from beginning to end. Once a few of these components have been completed, they prompt the writing of further components until you can complete a draft of the whole article.

Once you have that draft, you have an entirely new perspective of the article. No longer is the challenge to fill a blank screen or a clean sheet of paper but to correct or elaborate on material that is already there and to make it consistent and coherent with the material around it.

This is editing. Editing is much simpler than creating new material. At the editing stage, the material is in some sort of context and it is comparatively straightforward to check what it follows and where it is leading so that it can be modified with confidence. Modifications can be made in a much shorter time and with much less preparation than new text because they are usually discrete and are made in a context that is usually much clearer. Best of all, modern word processing makes the job of editing so much simpler than it was for our predecessors decades ago. Words, sentences or even whole paragraphs can be rearranged with a few key strokes and the result can be viewed instantaneously. It makes sense to plan to write in a way that takes advantage of the relatively new and valuable tools at our disposal.

The key, then, is to pass as quickly as possible from the writing or creating stage to the editing stage which appears, and is, far less complicated. Even if some of the initial editing is substantial and, in fact, involves creating and inserting several passages of text, this can usually be done within a framework that has a clear beginning and end, which makes the task much simpler. When editing, you can often make satisfying progress even in a few minutes of spare time during which you would not contemplate trying to write entirely new material. Your confidence builds and you get a feeling of having made good progress.

The difficulty of getting started is not simply a problem of thinking of what to say but the problem of being uncertain of how to say it. That uncertainty is needlessly heightened by at least five myths about scientific writing.

Myth 1. I must learn the special 'language' of research before I can write well.

There is a perception that there is a distinct language of science and research that has an idiosyncratic style that is formal, stilted and unlike the everyday language by which we normally communicate. Because it is unfamiliar to most people, it makes them uncertain about getting words together to get started. Fortunately, as we will see later, the perception of scientific writing as a stiff, formal and difficult medium is an illusion. It is perpetuated to a degree by the fact that it is easy to unearth examples of stiff, formal and difficult writing in the scientific journals. Not surprisingly, these examples are usually in articles that are also difficult to read. But, the basic language of science is simple, clear English—nothing more, nothing less. Certainly, many things discussed in scientific writing contain complex and, to many people, unfamiliar words and expressions because they describe complex and relatively unfamiliar things but the words that explain these things can be, and should be, disarmingly simple.

The style of English with which we are all most familiar and therefore most comfortable is the English we use in conversations and this is more than acceptable for a first draft of a scientific paper. And what's more, in later editing, we will not have to make major changes to that conversational style. So, we can dispense with the excuse that we can't get started because we are not used to writing in a 'scientific' style because, to all intents and purposes, there isn't one.

Myth 2. I must choose my journal before I start writing.

The editors of most scientific journals publish from time to time a *guide to authors*, a page or two that describe how certain things should be expressed, the units that must be used, the layout of references and other minor issues to make all articles in the journal consistent with what becomes known as the 'house style' for that journal. These guides are seldom comprehensive and, to scientists starting to write an article, can be more of a distraction than a help in the main task of outlining on paper the major purpose and findings of their research. It makes a lot more sense to use these guides to edit your work when nearing the end of your writing. More importantly, until you have done most of the 'thinking in words' that you need to develop your well-written and reasoned article, you are unlikely to be in a position to make a sensible choice of the appropriate journal anyway. It is more rational and far less distracting to commence your writing with an open mind about the journal and dedicate your concentration to what the article will contain and the order and logic in which it will be presented. Then, eventually, when you have a clear picture of exactly what you have to offer, the choice of journal and the consequent adjustments to the text to fit its style can be made without problem.

Myth 3. If English is not my first language, I will need help from the beginning.

It is true that English has become *de facto* the language of science and to write and publish in other languages restricts an author's potential readership to a small fraction of that when the article is in English. Almost all scientists, whatever their native tongue, have to learn English to learn what other scientists are doing and to be part of the scientific community in their field. But if English is your second or third language, you are probably not as familiar as native English speakers with the vocabulary and idioms of the language and therefore feel uncertain about writing freely without immediate help. Most of this insecurity is misplaced.

Science has a language of its own that has nothing to do with the scientist's native tongue. It is the language of logic in which reasoned arguments are developed from well-presented evidence and lead to sound and consistent conclusions. That language is the same regardless of the origin and preferred tongue of the person who writes it and good scientific writing depends primarily on expressing the science precisely and clearly. Subsequent editing by a native speaker to tidy up English expressions and comply with modern vernacular is relatively easy and the article will be a good one. If the expression of the science is poor, no amount of correction of the English can turn it into a satisfactory paper. In other words, a limited fluency in English is not a valid pretext for putting off writing an article to announce a good piece of research.

In fact, non-native English speakers often have unexpected advantages when it comes to writing science. In many English-speaking countries, schools are spending less time in teaching the basic grammar of the language. This results in a lot of native English-speaking scientists having real difficulties in recognising grammatically incorrect sentences or analysing why certain sentences don't seem to say what they want them to say. By contrast, the grammatical training in non-English-speaking schools is often more meticulous, and students learn not only to recognise verbs, nouns, adjectives, adverbs and prepositions but also how to use them effectively, albeit within a more limited vocabulary.

The very fact that the vocabulary is limited may also be helpful because it is usually accompanied by a limited knowledge of elaborate figures of speech and complex groupings of words. The resultant article will be restricted to words and terms that express thoughts plainly and economically. By a happy coincidence, this is precisely what is required for scientific writing so non-native English speakers often have an instant advantage.

Myth 4. I must write my paper sequentially from beginning to end to make it coherent.

This implies that you must have all of your information, all of your thoughts and all of your reasoning clearly developed before you put pen to paper. This is over-ambitious, and much valuable science lies languishing in filing cabinets because its creators have left it there waiting for the inspiration that will make it all clear to them. Writing is an integral part of the scientific process and the discipline of thinking about and writing some parts of your article almost always develops new perspectives for other parts of the article. So, accept that you are going to develop your paper in drafts, possibly starting with the easy bits and using them for inspiration for the harder bits. At the same time, you will be getting away from the difficult, 'filling-blank-pages' stage to the 'editing-existing-text' stage, which is far easier to manage, particularly in short periods of available time. Many people like to start with the *Materials and Methods* because they are descriptive and require little further interpretation. Others start with a draft of the *Results* to give them a base for further thinking. Some feel it is important to have a *Title* from the very beginning and that is fine, although we will see that this may not be the *Title* that remains after further editing. Others suggest writing the *Summary* first but, as we will see on page 49, it is much easier to write at the end. Whatever the order you choose for writing the first draft, you will find it much easier to integrate and rationalise the components by editing them later rather than attempting to do it in sequence as you go.

The style in which you write is, of course, important. But, at this stage, it is secondary to having a sound, logical and scientific structure to your article and this should be your goal in constructing the preliminary drafts. Sure, if you write well and with flair and fluency, so much the better. But, if you struggle to achieve flair and fluency to the extent that it distracts your concentration on achieving a sound, logical and scientific structure, then it will be counter-productive. It is much better to give your attention solely to constructing a well-structured draft at this stage and plan to edit it for style at a much later stage. This is a sensible plan for all writers of scientific articles but particularly for those whose first language is not English. They should not be inhibited by perceived or real difficulties in English that could hinder them from writing rigorous science and logic, a task in which they are likely to have no relative disadvantages. In fact, to express their reasoning exactly as they wish in complex segments it is preferable to write at least this part of the first draft in their primary language rather than to mess it up because the unfamiliar language has become a further obstacle.

Myth 5. I have negative results and editors won't publish results that are negative.

Many people complain that editors and reviewers discourage them from publishing results that are considered negative. They argue that unless treatments applied or data obtained demonstrate clear responses, the resulting paper is unattractive and difficult to get published. They argue further that, if negative results are not published then other researchers would be unaware of them and would therefore repeat the same research with the same negative result and the cycle would continue. This problem is considered serious enough in the field of ecology and environmental biology to have encouraged the appearance of a specialist journal, the *Journal of Negative Results*.

That seems to be an unnecessary response because a research project that is properly related to a hypothesis does not have to be concerned whether the results prove to be positive or negative. In other words, if there is strong scientific and logical backing to support an expectation that there should be a positive outcome and the results prove that there is no such outcome, this is a decisive result. Of course, it will require robust experimentation with adequate numbers and appropriate statistical backing to avoid 'type 2' errors. It should lead

THINKING ABOUT YOUR WRITING

to a vigorous discussion and it will be meaningful scientifically if it modifies conventional thinking in the field to accommodate the fact that an outcome that was logically expected to happen did not eventuate. A result that is negative because it contradicts an expected positive result is valuable and, if expressed in these terms, is unlikely to be rejected by referees or editors.

The problem with negative results is that they are often not presented in a way that emphasises their value. On the other hand, if results are negative because they have not been preceded with logical arguments that make it surprising that they are negative, or because the experiment has been poorly conceived and executed, then they do not deserve to be published. They will contribute little or nothing to the advancement of science.

This is yet another reason for developing an hypothesis in the introduction to any scientific article. The terms 'positive' and 'negative' are irrelevant when discussing the results of any experiment, survey or inquiry based on a well-reasoned hypothesis.

Writing about your thinking

WORDS ARE PRECIOUS IN WRITING SCIENTIFIC ARTICLES THAT inform other scientists of new work and ideas. The right words need to be in the right places for the right reasons if they are to do their job properly. This section illustrates how authors of scientific articles can manage the information in each part of their article to produce a rigorous, concise and readable paper, enjoyable for both writer and reader.

The *Title* 17

The *Introduction* 20

The *Materials and Methods* 28

The *Results* 31

The *Discussion* 39

The *Summary* or *Abstract* 49

The other bits 51

Editing for readability and style 55

When it is finished, your scientific article will or should have two very important attributes. It should have a sound logical and physical structure and it should be written in a style that is readable, precise, clear and brief. At the beginning, it has neither of these of course and it is a tall order to set yourself the task of getting both the style and the structure right simultaneously. Rather than being too ambitious, plan to do it in stages or drafts. It makes sense to write the article concentrating first on its structure and then editing it later to ensure that it meets the requirements for readability and style. I am suggesting this order on at least three grounds.

First, the need for a sound structure and the principles behind such a structure are not as well recognised as the need for a readable style. Indeed many people believe that a research paper is simply an exercise in English composition. So let us get it in perspective from the start.

Second, a sound structure is the product of sound scientific thinking and reasoning. This is your realm. It is the area in which you have had most training so use your skills to boost your confidence in writing. Then, later, think of the editing for readability as fine-tuning.

Third, science and reasoning know no language barriers; they are a language of their own. And this language is universal, regardless of the tongue in which you normally express yourself. If you are not a native English speaker you are not at a disadvantage relative to those who are. So, get the structure right first. Then, even if you need help later to modify the syntax or a few words, it should be only a minor exercise. By contrast, if you distract yourself and compromise the logical construction of the article by attempting to write flawless English from the beginning you will have little chance of completing a good piece of scientific writing. No amount of correction of the English can convert a poorly structured paper into a good one.

The physical structure of a scientific article is well known and, with a few minor variations or additions, is practically universal.

Title

Summary

Introduction

Methodology

Results (including tables and figures)

Discussion

Acknowledgements

Bibliography

In this section, we will examine what should go into each of these in the order in which they are usually found in a paper but this does not mean that, in practice, you must follow that order. In fact many people like to start with something that is relatively easy like the *Methodology* to ease them into their task. By contrast, the *Summary*, though it is physically the second heading, is much easier to write after you have completed the *Discussion*.

However, the logical structure of a scientific article is different from its physical structure and it revolves around the *Introduction*. So, it is a good idea to face up to this and begin crafting the *Introduction* as soon as you can.

The *Title*

The primary aim of writing a paper is to have it read. The *Title* is the first—and, in most cases, the last—a potential reader will ever see of your paper. For every person who reads at least a small part of your text, a hundred or more will probably have read your *Title* to help them decide whether or not to read on. So, it is worthwhile spending the time to craft your *Title* carefully.

It has two functions:

- to attract other researchers to read your paper and

- to provide the best information possible to help electronic search programs find your paper easily.

… you are in competition with every other author who had an article in the current issue for the reader's precious reading time.

Creating a title should therefore not merely be to provide a rough guide to the reader about the general field in which the work was carried out. Be under no misconception, in the *Contents* page of the journal you are in competition with every other author who had an article in the current issue for the reader's precious reading time. You have to produce something that is not only factually correct but which stands out from the mass of other titles on the *Contents* page or in the list of results from an electronic search.

With this in mind, look at the *Contents* section of the nearest journal you can find. You will be astonished by the number of titles that are unimaginative, uninformative and therefore unattractive. The most common of these will probably take the stereotyped form: *The effect of A on B (or The influence of A on B).*

A title like this gives little incentive to turn spontaneously to the body of the article to find out more. Worse still, it does not tell you what happened. 'A' may have affected (or influenced) 'B' by making it better, or worse, or it may not have changed it at all. How frustrating to read a paper in which the title announces that something is supposed to influence something else but the text shows that, in the end, there was no effect.

Other titles may provide more information, but is it information that matters? *A linear-based, retrospective clinical study of the incidence of Peabody's disease in a rural based teen-aged population …* or, catering for a popular fad to use colons as often as possible in titles: *The incidence of Peabody's disease in a rural based, teen-age population: a linear-based, retrospective clinical study.* The important question is whether this is information that is likely to entice a prospective reader to seek further. It makes a feature of the methodology but the findings remain a well-kept secret. If, in fact, the main thrust of the article was to show that this methodology is a new and exciting way of studying Peabody's disease, then there may be a case for this form of title. But if, as is likely, this is not the case, then the authors have wasted a great opportunity to 'sell' their article. In short, *Titles* that keep secret the contents of the paper and imply that you won't find them out until you read on, completely miss the point.

You can do much better than this, and here is a set of guidelines that helps you produce *Titles* that meet the two functions of persuading potential readers to read some more of your article and ensuring that search programs will find your article when appropriate words are keyed in.

... make sure that your title blurts out as much as possible of the research news that your article is going to talk about.

Carefully choose the key words in your article (these days, most editors ask you to do this anyway).

Rank these words in order of importance; if you were asked to summarise your article in one word, it would be your first key word!

Construct your *Title* using all of the key words and trying, as closely as you can, to put them in their rank order. This exploits the principle that the reader perceives that the words you use first in a *Title* are more important than those you use later. You will seldom be able to get all of the words in the exact rank order that you chose but, if you get close, you are likely to give the reader the same impression as you have of what is important in your paper.

If the *Title* is too long, drop off the least important key words first, but don't abandon them; you will need them to fill in the *Keyword* section later on.

Now, edit this draft *Title* to interpolate an indication of your main result or main conclusion—in other words, the real reason for writing the paper in the first place.

In summary, make sure that your *Title* blurts out as much as possible of the research news that your article is going to talk about. At first, you may think that this is revealing the plot too early and that readers won't be encouraged to read on if they know what is to come. It is quite the opposite. First, you can't reveal anything but the most general information in one short title. More important, what you do reveal begins the all-important task of developing the readers' expectations and providing a framework around which they will be better able to understand and retain the details of your article. Without that expectation, they are far less likely to bother reading further.

As an example consider the title, *The effect of an extract from* Leptospermum fasciculum *on wounds infected with* Staphylococcus aureus.

If the authors felt that the results of this study were the most important thing that their research had to announce they might edit their dull first draft by telling us about the main result: *Extracts from* Leptospermum fasciculum *reduce infection in wounds by* Staphylococcus aureus.

If, however, they thought that their main message would come from their conclusion then the title might be edited to: Leptospermum fasciculum *has the potential to replace conventional antibiotics in reducing infection in wounds by* Staphylococcus aureus.

The last two are clearly far more attractive and informative than the first. There are so many stereotypic titles in the literature that it is relatively easy to give yourself a competitive advantage by designing and editing a *Title* that tells readers what they can expect. Not only that, by being more detailed, you are likely to be less ambiguous and therefore more scientific. And, more information does not necessarily mean more words. For example:

The influence of season of calving on the performance of Holstein cows. The lack of information here raises at least two questions, 'What sort of influence—good or bad?', and 'What performance—running, jumping, producing milk, producing meat … ?' A better title might be *Holstein cows produce more milk if they calve in spring instead of autumn*. We have added one extra word but have increased the information and the impact disproportionately. Similarly, a title *The influence of manganese on petunia leaves* conveys much less and is less attractive than *Manganese brightens the colour of petunia leaves* and takes exactly the same number of words.

Some titles have words and phrases that appear to have been designed to put readers off. You should hesitate whenever you are tempted to use such words. Those that start 'Some aspects of … ', 'Observations on …' give the impression that even if we were to read the paper we would only get half a story. Words such as *influence, effect* or *change* are usually directionless and ought to be replaced with more helpful alternatives such as *reduce, increase,* or *brighten* which tell readers the direction of the influence, change or effect. In the same way, words like *correlation* or *relationship* in a title give readers no perception of what is happening unless they are referred to as being *positive* or *negative*.

As an exercise, look for some dull titles in a familiar journal and for each title, read the summary of the paper that follows. Using this information, compose a new title that is at least as brief, more specific and informative, and more helpful to a 'key word' system than the original. Such an exercise is not only good practice but it will quickly illustrate the poor standard of so many of the titles currently being used.

When drafting your list of key words, make sure that they are as effective as possible by making them as specific as possible to your article. If, in the experiment described in your article, you found a large and significant correlation between A and B, by all means use A and B as key words but ignore words like *correlation* (positive or otherwise) which seekers are hardly likely to tap into the search program. For the most part, key words should be nouns, so adjectives like *large* and *significant* are also not going to be very helpful.

Search programs never search only for those words in the *Key words* section. They always include the *Title* in the search as well. In fact, some modern search programs use the power of current computers to search the whole article, so the *Key words* section may become less important in the future. Nevertheless, even with the power of modern computers, the *Key words* section can be a useful tool to restrict the limits of a search. It is a waste to repeat in the *Key words* section, words that you have already used in the *Title* but be sure to include those that you may have excluded when you edited your *Title.*

The *Introduction*

A scientific paper communicates new research information to scientists, so its first objective ought to be to demonstrate that the story being told is worth telling—that it is scientifically sound and addresses an issue plausibly and logically. The *Introduction* is where the author convinces the reader that the work has been well thought out and, at the same time, orientates the reader's thinking along the same pathway as that of the author. In short, it is the powerhouse of your article that should be feeding life into every other section of the paper.

A good *Introduction* goes much further than just stating the problem and acquainting the reader with the relevant literature. It should describe a series of logical steps that end in a statement of what the experiment is about, why you did it and what you expected to get from it. If you have done a good job in constructing the *Introduction,* you will have converted the reader from a passive and relatively disinterested recipient of the information you want to communicate into an enthusiastic seeker of information.

Most texts and guides to scientific writing give general instructions to authors about how to write an *Introduction.* Most are not very specific and therefore don't help much. A random selection from some of these texts on what you should do to write an *Introduction* includes:

Define the scope of the study

Define the problem

State the objective

Identify gaps in the knowledge about the subject

State the purpose of the experiment

Summarise the background to the research
  (sufficiently but not too widely!)

State the question that you asked

Provide a context for your investigation

Briefly explain the theory involved

Present an hypothesis or an expectation.

You should, of course, do all of these things, but the question is, 'How?' and advice on this is very rare. To do each of these points justice could take five, ten or more pages and most scientific journals do not give you this luxury. The technique of writing an *Introduction* effectively starts with understanding two relatively simple principles.

The first is that the hypothesis is the key to the *Introduction*. The second is that by justifying the hypothesis logically and scientifically, you provide just about everything necessary for readers to understand what your paper is about and why you wrote it. In other words, it automatically and concisely covers all of the points above in one clearly focused statement.

So, a good *Introduction* consists of two distinct sections, a short statement of what the author could logically have expected to find before starting the research, preceded by a reasonable scientific proposal justifying that statement. These two sections in many well-written papers are often embodied in two paragraphs or, if they are very uncomplicated, within a single paragraph containing the two sections. The second section is shorter than the first and contains the hypothesis. The first section has no other purpose than to justify the hypothesis. As a result, the technique is remarkably simple. The *Introduction* so produced is also simple and relatively concise but—be careful—the thinking, reasoning and groundwork behind it may be far from easy. However, if you are prepared to work through the process of thinking about and writing the logical justification of your hypothesis, you will be rewarded by finding that not only will your *Introduction* be simpler to write but every other section of the rest of the paper will also be simplified because it will have a central focus. Most important of all, readers will interpret more easily what you have to say in the rest of the paper because your *Introduction* will stimulate them to anticipate certain information and you will be presenting it to them when they expect it. This, in turn, will mean that they will be more likely to interpret your information in the same way as you do and be less likely to be confused by it.

The secret is undoubtedly the expression of a logical hypothesis. But not all research papers are about the results of experiments in which treatments were applied specifically to test hypotheses. Some present the results of surveys of new material or new areas. Some report observations on populations of people, plants, animals, landscapes or compounds to which no treatments were applied. Some test new techniques for measuring or observing without applying treatments. How can you formulate sensible hypotheses in these cases? At first, it may seem quite difficult, as I realised after being confronted by researchers with this issue over many years. When stressing the desirability of having an hypothesis as the intellectual base of their article, I was told continually that their experiment was different or was not really an experiment at all and therefore did not have an hypothesis. They tried to justify this belief with a variety of reasons:

- I was just looking for base-data from which an hypothesis might be formulated at some later stage. I had no idea what I might find in this initial study.

- Our laboratory bought a new instrument that measures things that we couldn't measure previously and so we used it to see what was there.

- Mine was not an experiment but a questionnaire. How could I possibly hypothesise what answers I would be likely to get from respondents to a questionnaire?

- We were just trying out a new methodology but we did not know whether it would be better or worse than the old. The results were not important, only the technique.

- I inherited the experiment that I want to describe from someone who has since left the research organisation (or was promoted to administration, or died) so I have no idea what original hypothesis they might have had.

- I have expressed the aim of the experiment I am about to describe so I don't need an hypothesis, do I?

Sadly, in all of these cases the authors were inferring, probably inadvertently, that theirs was an article about how they groped about in the dark, hoping that something worthwhile might turn up. But, in none of them did the circumstances give them the excuse to avoid predicting what might logically be expected as an outcome, in other words, an hypothesis. Think about readers confronting the article for the first time. They are searching for some scientific incentive in the *Introduction* to make them continue reading further. Can you imagine a statement expressing any of the excuses above filling them with any inspiration to persist? More likely, they will conclude that the author was fumbling about without thought or reason and will look for a more predictable article with which to occupy their precious time. Fortunately, in each of these instances there is always a latent hypothesis which, when expressed, gives both the author and the reader a focus for the rest of the article. Let us look at each of them in some detail.

I was just gathering data.

Nobody can afford the luxury of just gathering data. Of the infinite number of things in the world that could have been investigated and the infinite number of ways of doing the research, you chose only one of each. Your job in the *Introduction* is to convince the reader that you have chosen sensibly on both counts. That means explaining why you chose the particular topic to study. It means explaining how eventually you plan to use the 'base data'. It is virtually impossible to do any research without an objective and, in just about every case, when you have an objective you have some idea of what you might find. Otherwise, how can you justify collecting certain sorts of data and not others or how can you explain rationally why you used a certain methodology to gather those data? The most satisfactory way to address these issues is to progress beyond the objective and predict what you are going to find, then justify that prediction. Then, it will make sense to the reader why you did not measure certain things but concentrated on others. Later in the article, both you and the reader will be able to relate the quality of the data you gathered to the purpose for which you gathered it. It will probably even help you explain the experimental design you used. Most importantly, you will have built a strong focus around which readers can begin to anticipate what they might find in your *Results* and, later, in your *Discussion* and that will give them the incentive they need to continue reading your article.

We have just bought a new instrument.

Instruments are tools that allow you to do research. They are important to the description of the research only when a researcher uses them cleverly or has a particular purpose in mind when choosing that instrument. You will let yourself down as a scientist if you gave the impression that the availability of an instrument was sufficient justification for the research that you are about to describe. The clearest and most effective way to ensure that the reader recognises the scientific purpose of the research and, if relevant, your cleverness in using the new instrument to carry it out is to describe what you expected to happen—your hypothesis.

It is a questionnaire, not an experiment.

Questionnaires are a legitimate and often highly effective tool for carrying out research in the social sciences. Their effectiveness is normally directly related to the amount of thought and planning that goes into the composition of the questions. These well-planned questions are invariably the result of anticipating the likely answers. This enables the questions to be fine-tuned to ensure as high a proportion of meaningful responses as possible from as few questions as possible. In other words, the questioner has to have a reasoned hypothesis about how people will respond. An account of that hypothesis and the reasoning behind it in the *Introduction* is the ideal way to explain your logic and prepare the reader to grasp and understand the rest of the article.

This is methodology, not experimentation.

Declaring an article to be a methodological paper is not an excuse for avoiding a statement of the hypothesis. If the experimenter did not think that a certain methodology was better in some way than another and did not have good reasons for thinking so, then there would have been little justification in doing an experiment to test it. In other words, it is possible to have an hypothesis about methodology. For example: *Method A is better than method B for a particular purpose.* Or: *This new method will enable me to study something that was impossible to study satisfactorily before.* Or perhaps: *This new method will be just as precise as the old but will be less expensive.* In every case, the hypothesis on which the experiment was based is the ideal medium for preparing the reader and consolidating the theme for the whole article.

It was an 'inherited' experiment.

Writers who did not have an hypothesis when they became involved with the experiment are sometimes reluctant to create one to introduce the article that describes that experiment. The original lack of an hypothesis may have been because they, themselves, were unprofessional at the outset or because they took over an experiment that was designed by someone else. They must remember that, regardless of their thinking, or lack of thinking, during the course of the experiment, they are engaged in telling a scientific story as clearly and as succinctly as possible. They cannot do this by making the reader as confused and disoriented as they, or the person from whom they inherited the experiment, may have been. So, if the results that they are about to present and discuss do, after reflection, shed light on an hypothesis that was conceived sometime after they were obtained, it is important to give the reader the benefit of that reflection and present the hypothesis accordingly. The experiment may not have been done deliberately to test the hypothesis but, in reality, it did. Readers want to know what you, the writer, might have expected to find and be able to assess every sentence of the rest of the paper against this expectation. With a well-reasoned hypothesis in front of them, they can now read ahead in anticipation instead of thrashing about trying to find out what it was all about.

The aim will do.

An aim (or an objective) is very different from an hypothesis and is no substitute for one. It neither stimulates the expectation of the reader nor helps the writer produce a tight, coherent article. An aim states what you intend to do, and there is nothing wrong with that, but it does not specify why you intend to do it. For example, the aim of an experiment could be 'to compare the body mass index of a sample of children from a rural population with that of a sample of the same age from a city population'. A statement of the aim does not need to be justified. By contrast, a stated hypothesis for the same study might be 'Children in the city have more access to junk food outlets than their counterparts in rural areas. We reasoned that, if access to junk food is a cause of obesity in children, rural children would have a lower body mass index than city children.' In this case the statement had to be justified before it made sense. The *aim* is therefore much easier to formulate than the *hypothesis*. In fact, the formulation of an hypothesis is a major intellectual exercise. Cynically, one might suggest that intellectual laziness is why some people try to think up pretexts for avoiding having to present an hypothesis in the *Introduction* to a scientific article. However, if it is well formulated, it makes it comparatively easy to write the rest of the article. More important, it makes the rest of the article easier for the reader to follow and interpret.

... the soundness of the structure of your article depends on a well-reasoned and clearly stated hypothesis.

So, there is little doubt that no other single statement in your paper is as important as the hypothesis. Despite this, you are not compelled to express it in the stereotypic form, 'The experiment tested the hypothesis that ... ' In fact, if you have some objection to using the word 'hypothesis' you do not even have to use it at all. It is a 'prediction' or an 'expectation' and you can use these words or their stems if you like because they convey the same sense. You can also use expressions like 'we reasoned (or deduced) that if we applied "A", we would get "B"' as is illustrated in the example above about obese children.

Whatever you call it, an hypothesis, an expectation, a prediction or a piece of reasoning, the soundness of the structure of your article depends on a well-reasoned and clearly stated hypothesis. It becomes the focus for the article and, so that you don't lose that focus, it is a good idea to write it out in red ink, in capital letters, or whatever method will emphasise it most, and pin it up in front of you. You, the writer, and the reader, are now both fixed on the same objective but the reader will probably finish the paper in 10 minutes and is unlikely to forget the objective. It may take you several weeks or months to write your paper among all of your other tasks, so a constant reminder of your hypothesis will help keep you on the right path.

Having stressed the vital role of the hypothesis, I have to stress also that the hypothesis itself is only a summary or a conclusion derived from the reasoning that underpins it. You will often see articles in which the authors seem to have inserted the hypothesis because they have been told that they should. But, it is the clarity and persuasiveness of the reasoning that orientate the reader, not just the hypothesis. So, an *Introduction* that is rambling and discursive but ends in a statement of a hypothesis that has not been fully and clearly justified, is not much more helpful than one with no hypothesis at all.

What we must avoid is a statement that says something was attempted in a certain way to 'see what happened', or 'it seemed of interest to examine this phenomenon further', or 'there are no reports in the literature of a study of this, so one is presented here'. All of these convey a randomness of thought and a lack of scientific discipline that signal to the reader that the next few pages of text are going to be hard to read. When you read these sorts of statements you will probably read results and discussions about almost any observations that come to mind so long as they conform to the general area that was so vaguely defined.

The reasoning behind the hypothesis—the other part of the *Introduction*

Now that the last part of the *Introduction* has been decided, the first part can be filled in. Many people have difficulty in deciding what should go into this part, what work should be quoted and what should be left to the *Discussion* or even left out altogether.

The decision is easy. Only that material forming part of the logical series of statements leading to your hypothesis has a place in your *Introduction*. Just because you know of work in the general area, or because some well-known scientists in the field might become annoyed if they are not quoted is no rationale for including them in the *Introduction*. This is not the place for dropping names or doing favours for other authors by adding to their numbers of citations. Only if they contribute to the development of your logic should they be included.

In setting up the hypothesis, remember that a result that supports your hypothesis does not mean that the hypothesis is infallible. For example, if you were to set up an experiment to test a generalisation and found that your results fitted the hypothesis, this is no more than additional evidence that it is a good explanation of the phenomenon that you were examining. Another series of observations made under slightly different circumstances might fall outside the generalisation you have made and prove the hypothesis wrong. So, be careful with your wording. You can support an hypothesis but never prove one.

On the other hand, if your observations caused you to reject the hypothesis—and, of course, your experimental methods were sound—then you can be much more positive about your conclusion and you could say that your results *disprove* an hypothesis. For this reason, and for the sake of the written story, it is often more convenient if the prior evidence happens to allow you to frame a proposition about the known information in a way that your results may reject.

I stress here the necessity to arrange your arguments and your hypotheses in a logical, precise order. Everyone knows that even brilliant scientists—in fact, especially brilliant scientists—

do not always think logically and precisely. Many of the great discoveries of science have developed from flashes of brilliance that often came under unusual circumstances. Folklore has it that Newton developed his laws of gravity after being struck with an apple while relaxing under a tree. His brilliance lay in being able to relate a 'happy accident' to the known facts and then build up the laws of gravity. Fleming had his stroke of luck when, in apparently substandard working conditions, a plate of culture medium became contaminated. He, too, had the brilliance to think through the consequences of this, until he and Florey isolated penicillin. Parkes and Polge discovered that germ cells can be protected from freezing temperatures, because a technician made a mistake and mixed glycerol with some samples of semen which they were attempting to preserve. These scientists also worked carefully in the reverse direction from the result to the reason for it, to achieve a major scientific breakthrough.

You are therefore obliged, out of respect for your readers, to give them only the distilled essence of your thought processes.

Down at our level, many of our ideas also come from inspirations. Fortunately, there is no way that we can train ourselves only to develop ideas logically. If we could, we could leave the whole of scientific discovery to computers. Most ideas do not stand up to testing against the facts we find in the literature or through experimentation. Sometimes, we set up an experiment to test what seems to be a good hypothesis at the time, only to find that our techniques are inadequate or our ideas were not so smart after all. Nevertheless, such experiments sometimes yield interesting data that could provide valuable information—but, for a different hypothesis from the one we were originally testing.

If the processes of scientific thought are so haphazard, why am I suggesting that you set them out so logically when writing a paper? A basic rule of science, after all, is that we should be scrupulously honest. Shouldn't we record our ideas, discoveries and failures in chronological order? If we tested an hypothesis that we have now scrapped but, in so doing, saw how the results spread new light on a different hypothesis, shouldn't we say so?

The answer is, 'Probably, no!' You have had anything from six months to, perhaps, 20 years to test, reject, re-form and re-reject ideas and hypotheses. You have slept, eaten and worked with them and in the end you have come up with what seems an important piece of information. Your readers, on the other hand, have about a minute to cover the same ground. You are therefore obliged, out of respect for your readers, to give them only the distilled essence of your thought processes. Doing scientific research, and writing about it afterwards are not the same thing. They have very different objectives. Research is the finding of new information by testing hypotheses, rejecting or accepting them, refining and re-testing them, and finally coming up with new knowledge. Writing is the dispassionate recording of the knowledge in a manner that presents the data in an honest, plausible and straightforward way. The blind alleys you traversed, the disappointments, and the poor techniques along the way cannot be allowed to impair the reader's chances of seeing your new information as a clear-cut piece of reasoning. Therefore, in the *Introduction*, you should present only the

SCIENTIFIC WRITING = THINKING IN WORDS

hypothesis (or hypotheses) that you intend to test and discuss in your paper and you should present only that supporting information that makes these hypotheses sensible.

The net result is that, far from being a loose preamble, an *Introduction* is a very tight, clearly defined piece of writing the moment that you settle on the final form of your hypothesis. It is not a general review of the literature for people new to your field of study and it does not need to begin with broad, sweeping statements, largely irrelevant to the experiment, before getting down to the main business of making the hypothesis a plausible statement of the phenomenon you were studying. It is often shorter and more focused than you find in many scientific papers and, in most cases, this is a good thing. However, many authors, and occasionally, some editors, feel the need to place the work in context—how they came to be doing work in this field in the first place, or where it fits in the bigger picture, or what pragmatic problem led to the need for a scientific experiment to find an acceptable solution. An *Introduction* consisting of an hypothesis and its justification and nothing else, sometimes does not satisfy such a need. The challenge is to do this in no more than a few sentences at most. Many introductions are made unnecessarily long by providing 'comfortable' statements of background in the misguided belief that these ease the reader into a receptive mood for understanding the detailed subject of the experiment. The problem in describing the wider context of the work is that many authors cannot decide how far from their experiment to start the statement of its background. For example, an experiment into an aspect of the physiology of flowering in tobacco does not have to be introduced by a statement of the economic value of the tobacco industry. A study of antibiotics on bacteria affecting the respiratory tract of humans is not necessarily made clearer by a general discourse about the statistics of respiratory diseases on a country or world basis. On the contrary, this broad level of *Introduction* may be a distraction unless it has a either a direct bearing on the logic of the hypothesis or, occasionally, is a preamble to broader issues to be taken up later in the *Discussion*. The tightening of the scope of the *Introduction* almost inevitably means that it is shorter that it might otherwise be, but this is generally appreciated by editors seeking to economise on numbers of pages per article. It is certainly appreciated by busy readers who are always anxious to get to the new material that the article might offer.

… far from being a loose preamble, an Introduction *is a very tight, clearly defined piece of writing.*

There is an important principle here about describing 'context' in *Introductions*. If the contextual information you are proposing is going to be used in the *Discussion* or supports the hypothesis or justifies the methodology, by all means incorporate it into the *Introduction*. If not, leave it out because it can only be a distraction that will swell the size of your paper without enhancing its scientific validity. In any case, aim to incorporate 'context' in a sentence or two as part of the more important task of justifying your hypothesis rather than according it a status in its own right by presenting it in one, or even more, paragraphs before introducing your work.

The *Materials and Methods*

The *Materials and Methods* is a great place to get your writing under way and build up some confidence for the more challenging parts of your paper. It is generally uncomplicated because it does not require that you do much interpretation. What you did was what you did and you can't change that—even if, on reflection or with the benefit of experience, you think you could have done it better. The task here is to describe what you did in such a way that an informed colleague in the field could repeat the experiment based on the information you provide.

The task here is to describe what you did in such a way that an informed colleague in the field could repeat the experiment based on the information you provide.

Despite its relative simplicity, there can be pitfalls in writing a good *Materials and Methods* that stem from your over-familiarity with what you are describing. The skill is in knowing what can be left out but also what *must* be included to allow a reader, unfamiliar with the work, to follow it and, possibly, repeat it. The section can get out of hand if you try to include too much detail so take a hard look at what you *could* describe and see what you could remove without reducing its accuracy or clarity. On the other hand, you may describe a technique with which you are almost contemptuously familiar and, quite unconsciously, leave out important details that would leave a potential reader floundering. That is why you need the view of an outsider. In the end, the best way of checking is to apply the test literally and put the completed *Materials and Methods* in front of a competent colleague who was not directly involved with your study. Ask if they could, in fact, repeat the experiment using the information you have given them.

But, in the meantime, there are some principles that can help your writing of this section.

Paramount among these is the part of your article that is most likely to be 'skimmed' by impatient readers anxious to get quickly to the *Results* and the *Discussion*. In fact, recognising this, some journals are now relegating the *Materials and Methods* section to a sort of appendix at the end of the article and presenting it in smaller font. However, wherever the *Materials and Methods* is placed in your chosen journal, some readers may return later to scrutinise it very carefully if the work described in the rest of the article interests them. Why not help readers to get a general idea of the way the experiment was done the first time around by ensuring that they pick up the essentials? You can do this by breaking the section into a sequence of meaningful subheadings that summarise the main features.

For example, if you were to see either of the two sequences of sub-headings below after having read a suitable introduction, you should have enough information to take in the details of *Results* and the *Discussion* that follow.

1.	2.
Experimental design	Design of survey
Experimental animals and diet	Selection of patients
Administering infusions	Information sought from patients a) before treatment
Sampling	b) after treatment
Analytical methods	Clinical information from doctors
Statistical analysis	Statistical analysis

It is a pity that this section is called *Materials and Methods* rather than *Methods and Materials* because until readers have an idea of your methods, the materials that you used will not mean very much to them. So, as the examples suggest, a subheading called *Experimental Design* or something similar describing how the experiment was done ought to be one of the first of your subheadings. Beyond that, there is no set sequence but, if you jot down all that you believe should be included, a sensible sequence will often become apparent. Ask yourself whether these headings only, in conjunction with the *Introduction*, could give readers a broad idea of how you went about the work. For example, in the second example they could envisage that you carried out a survey of patients that you selected using certain criteria, then got information from them before and after they received clinical treatment and compared this information with their clinical history obtained from their doctors. The purpose of, and the expectation from this experimental protocol would be already known, of course, from the *Introduction*.

The criterion that a competent colleague could repeat the experiment after reading your *Materials and Methods* section does not mean that you have to put all of the information in front of readers in this section. It means that you have to present them with a 'paper trail' that they can follow. Where you used novel techniques, or new modifications to old techniques, you must, of course, describe them fully and exactly. If, on the other hand, you used techniques that have already been described fully, then it is adequate and desirable to refer to the paper where the technique was first (or best) explained. Be careful though, to give credit where it is due. To refer to a recent paper in which the technique was used rather than to an older one in which it was originally described, is not only discourteous to the original author but fails to put the technique into historical perspective and does not enable a reader to proceed directly to a full description of it. Nonetheless, the original reference may be difficult to access or have been modified slightly since it was published. In this case, you could also refer to a more recent article and give the reader access to the complete technique while still apportioning credit where it is due. By using relevant references that readers would have to look up to be able to repeat the experiment, it is possible to describe relatively complex experiments with a lot of routine analyses in just a few lines of text.

Sometimes little things can be very important in giving a full and reproducible description of an experiment and they must be included. But resist the temptation to include every little detail just to be sure. Instead, make a conscious judgement about whether these details are really relevant to the experiment you are describing based on whether they would be necessary for someone to repeat the procedure. For example, a field experiment into the growth of plants might need to include information about the location, climate, soil type and rainfall in which the plants were grown. Most people would expect these to influence the results of the experiment, which would have to be interpreted accordingly. If, on the other hand, the experiment had been conducted in a light- and climate-controlled greenhouse, the experiment could theoretically be repeated in the middle of the Sahara desert or at the North Pole, so that where it was done and the details of the climate would be unnecessary and could kill a paper by drowning it in irrelevancies.

Sometimes little things can be very important in giving a full and reproducible description of an experiment and they must be included. But resist the temptation to include every little detail just to be sure.

Occasionally, new techniques are described for the first time in a paper and these often have to be validated to justify their use. Many people have trouble deciding whether this validation should accompany the description of the technique in the *Materials and Methods* section or whether it should be reserved for the *Results*. This becomes more complicated when the paper is more about the new technology than about the data that this methodology generated. The solution to the dilemma lies in the wording of the hypothesis. If the hypothesis clearly indicates that you predicted that the new methodology would open up new possibilities or would give clearer/cheaper/more precise information than another technology, then this is a paper about methodology. The validation and other testing that you did are clearly part of the test of the hypothesis and are much more appropriate in the *Results* than in the *Materials and Methods*. On the other hand, if your hypothesis proposed that the data obtained by using the technology would enable you to test a biological or sociological supposition, then the validation of the methodology is not the key factor in your article and is more at home in your *Materials and Methods*. In fact, if they were to be presented in the *Results* section before the important results, they could obscure the true value of these results.

It is often appropriate to describe statistical techniques in the *Materials and Methods*. The science of statistics has advanced greatly in the last 40 years and the use of statistical analysis is almost mandatory in quantitative disciplines these days. Once, when methods were unavailable or poorly understood and there was little or no computational power to do the calculations, the most obvious statistical manipulations of data were described in great detail. Nowadays, statistical analyses are considered, like chemical analyses, as part of the research worker's tools of trade, not the finished product. In most cases they are no longer rare enough to warrant special mention, certainly not in great detail. If you merely carried out a standard procedure like an analysis of variance, 't' test, or *chi* square, then simply say so. If the technique is more 'off beat' but well described in a published

SCIENTIFIC WRITING = THINKING IN WORDS

paper or standard text, then a reference to the source would be sufficient. Only if you have performed some original mathematical gymnastics do you need to describe them in depth in *Materials and Methods*.

A detailed description of the statistical methods you used may not be necessary but there are a few general rules to keep in mind. Most statistical methods used in publications are designed to look for differences between groups, treatments or across time, without considering the direction of this difference. Therefore the value of the probability given by the test is said to be bilateral or two-tailed. As a rule, when nothing is specified, probabilities are bilateral. But, if the direction of the difference is predicted by the hypothesis, unilateral probabilities may be used. If you are using an hypothesis in which the direction of the difference is predicted and you used unilateral probabilities, it is therefore important to specify it in the *Materials and Methods*.

The *Results*

To borrow a legal phrase, *Results* means: the results, all the results and nothing but the results. This seems so simple and obvious that you might consider it unnecessary to express it. But you would be surprised at the number of times that one finds results appearing for the first time in the *Discussion* of drafts of articles or, even worse, in the *Summary*. The *Results* section is where readers expect to find all the results that you intend presenting and so that is where you should put them. A reader who finds them anywhere else is likely to be confused about exactly what you found.

Separating Results *from* Discussion *preserves the objectivity of the* Results *which should be presented clearly and clinically without comment.*

Occasionally, in very short or very simple papers, there is little or nothing to discuss after the results have been presented. In these relatively rare cases, it may be possible to add whatever discussion is necessary to the results wherever it seems appropriate and the *Results* section would then be renamed *Results and Discussion*. Some journals, but not many, allow this but the practice is not very common or even very sensible. In fact, if the *Introduction* has been properly structured and the reasons for doing the experiment have been clearly stated, you will be compelled to discuss how the results met your expectations. That discussion is almost always neater and simpler to get across if it is in a separate section. Except for the shortest papers, a mixture of results and discussion always invites chaos in how the arguments flow.

Separating *Results* from *Discussion* preserves the objectivity of the *Results,* which should be presented clearly and clinically without comment. This encourages readers to draw their own conclusions and judgements which, no doubt, they will compare with yours later when they reach the *Discussion*. To qualify each result as it is presented with your own comments and comparisons gives the strong impression that you are trying to influence the objective judgement of readers before they have had the opportunity to see the complete picture.

Apart from the need to be scrupulously objective in presenting your results, there are two other, very strong reasons for separating the *Results* and *Discussion*. The first is that you are almost certainly going to be comparing your results in some way with those of others. This means that, if you mix *Results* and *Discussion* in the one segment, there will be two sets of information, yours and theirs, and this increases the danger of the reader confusing one with the other, on seeing them both and trying to absorb them for the first time. If your results—all of your results—are clearly quarantined in their own section, the reader can have no doubt that all of them are yours. The second is that it is very unlikely that you can logically discuss one isolated section of your *Results* without involving some of the other sections. If later sections have not been mentioned when you are discussing earlier sections, the outcome will inevitably be a chaotic blend of information that will confuse reader and writer alike.

What to present

… no one is more appropriate than you to determine the relative importance of each piece of the total information that you have to present.

In themselves, your results are not usually the most important new knowledge you are presenting to the world. It is more likely that it will be your interpretation of the results for which you will eventually be remembered and quoted. So, it is essential to have your *Discussion* in mind when composing your *Results* section. In all but the most exact disciplines, if you, or anyone else, were to repeat the research, you would not expect the treatments or observations to yield exactly the same numbers, but you would certainly expect that the new data could be interpreted in the same way. So, when wondering how to get your results from notebooks and work books, remember that it is not a matter of serving up large helpings of figures to the reader in endless rows and columns. Instead, you should present readers with information that you have carefully chosen and distilled to enable readers to understand your interpretation which will follow in the *Discussion*. Your raw data are often voluminous enough to occupy the space of two or three papers which means that the successful construction of the *Results* depends on your choice of the material to present and your decisions on how to present it.

It is a common characteristic of all experiments that some data are more important than others and no one is more appropriate than you to determine the relative importance of each piece of the total information that you have to present. Having done so, you can then decide on a strategy to present your information to readers in a way that conveys to them the same concept of relative importance. If you don't, they are likely to gain a wrong impression of what is important and what is not important among your data and, eventually, misinterpret the scientific story that you are telling them.

But, how do you, the writer, judge this importance and how do you get it across to the reader? The answer lies in the *Introduction* and particularly, the hypothesis that you proposed and justified within it. These gave the reader an insight about what you anticipated might happen in your experiment. Now, in the *Results*, you are going to present the data that might confirm or refute the outcome you anticipated. This is what readers expect to find and to present anything else would both disappoint and confuse them.

To refine this concept further, it is a good idea to assign a level of priority to all of the sets of data that you are considering for presentation. This will also clarify in your own mind the relative importance of your experimental data. This can be done by assessing each set of data according to its relevance to your hypothesis. Practically, four categories should be enough to allow you to do this. In summary, these are:

Category 1. Results that are clear and relevant to what you want to say about the hypothesis

Category 2. Results that allow you to say something relevant about the hypothesis but that are less convincing than the results in category 1

Category 3. Results that are interesting, substantial and are worth presenting but they don't have anything to do with the hypothesis, and

Category 4. Results that are not convincing and don't have anything to do with the hypothesis.

Now, with this information about the relative importance of the various components of your data, you have what you need to construct the *Results* in a way that makes it very clear which of your results are the important ones. First, do not use any of the results that you classified as category 4 and then, as much as possible, present the results in the order: category 1 before category 2, and category 2 before category 3. You can reinforce this weighting by using the text in the *Results* to emphasise the important information in your tables or graphs to which you particularly wish the reader to pay heed.

Sometimes data that are classified as category 3 need careful reflection. We all know that data like this crop up from time to time. They are interesting and shed light on a phenomenon that you were not directly looking for or, in other words, was not part of your hypothesis. To feature it as if it were the main part of the *Results* would be somewhat puzzling to a reader who had been following the logic of your thinking to this stage. But, to leave it out because it did not relate directly to your hypothesis would be a pity and a waste of interesting information.

Admittedly, the hypothesis says what the paper is about, so results that do not relate to that hypothesis should clearly not be the main feature of your paper. If you decide that the exception to this is so important that it must be given a lot of exposure, your paper will begin to lose its focus and direction. When exceptions take over the paper, it is time to reconsider your results. Perhaps, if you have too much material that is worthwhile but unrelated to your hypothesis, you should think about writing about it in another paper. Alternatively, you may decide to abandon the paper in its present form and present the material under a different hypothesis for which these data are a suitable test.

Authors sometimes find it tough to cull ruthlessly results that they worked hard to collect but which they honestly have to admit are most correctly placed in category 4. This is hard to do because it means that they have to acknowledge having probably worked fruitlessly in that component of the experiment. Unfortunately, readers are not interested in how much work authors do. They want to know what authors found that was useful. In fact, if readers are overwhelmed with data that are boring and which they finally consider irrelevant, their impression will be that the author was unable to discriminate and that is not a quality that enhances reputations. It is a hard fact of scientific life that heavy pruning is a normal part of paper writing. If you find yourself reluctant to throw out irrelevant data, soften the blow on yourself by thinking of the damage it might do to your better data by simply diluting it and increasing the risk that a reader may miss your main information altogether.

What form of presentation? Tables, figures or text?

Unfortunately, readers are not interested in how much work authors do. They want to know what authors found that was useful.

Having selected the material that you are going to use and determined its priority, you now have to display it logically and concisely. Most data require some treatment. This may vary from complicated statistical analysis to simple tabulation of results and the calculation of a few means. It is always worthwhile attempting some alternative methods of treatment and presentation before deciding on which is the most suitable. At this stage you should be forming, in your mind, the arguments that will eventually become the backbone of your *Discussion*. This, too, is a process of trial and error to find the best alternatives for final presentation. Remember, you will be referring back to certain highlights of your *Results* for the basis of the *Discussion*. So, it is essential that the important points are made clear to the reader in the *Results* section. Whatever your final presentation—graphs, tables or text—you should arrange it so that the key information and key figures are in prominent positions.

But, first, let us establish three important ground rules:

- The *Results* section usually contains both text, and tables or figures. It *may* contain only text but *never* only tables or figures. You are compelled to describe your results in text, not just to present numbers.

- Tables (or graphs) and text should both be 'self supporting'. In other words, readers should not have to read the text to understand what a table is presenting. Neither should they be compelled to read a table (or figure) to discern what the text is about.

- Results presented in tables (or figures) should not be repeated verbatim in the text. Nor should the same results be presented in both tables and figures in the same article.

So, we need to know what should be put in tables and what should be written in the text so as not to be repetitious and boring.

When more than a few numbers are involved they are always difficult to read in a horizontal line of text. So tables that line up the numbers neatly in columns or rows, or graphs that illustrate trends are usually far more appropriate than text.

SCIENTIFIC WRITING = THINKING IN WORDS

But, remember that using tables and figures does not absolve you from the responsibility of making the text a coherent story. This does not mean that the text should present the same data as the tables. In fact, the text and the tables in a *Results* section have distinctly different jobs to do and each complements the other. That difference should always be at the front of your mind when structuring the content of the *Results* section because it helps you meet the commitment that characterises all scientific writing: to be precise and clear. In short, tables or figures are the means by which authors can ensure that they meet their obligation for precision and, having met that obligation, they can concentrate, in the text, on the second obligation, clarity.

Tables and figures are made up of numbers, so any desired but, of course, sensible level of precision is possible. The text gives you the opportunity to clarify and reinforce those aspects of the tables that will be particularly important when you come to the *Discussion* later. It is unlikely that every number in every row and column of a table is as important as every other one. In the text, an author can clarify for the reader the key issues in the tables by drawing attention to only those parts of the table that contain the important data. The text can also be very effective in drawing attention to patterns within, and between, groups of numbers and clarifying immediately the 'big picture' that they present. The numbers that are not of interest may be there for completeness and for readers who may want to see them but, by disregarding them in the text, the author can clearly signal that they are not going to play a major part in the scientific story being told and discussed. Even when describing the important data, the author can use the text more flamboyantly to say, for example, that 'treatment A was nearly twice as effective as treatment B in controlling the disease'. Such an imprecise and otherwise loose statement would be unthinkable as a piece of scientific writing without the precise numbers in the table to back it up. But, it gets across the main message with an emphasis that the mere numbers alone could not have done.

... tables or figures are the means by which authors can ensure that they meet their obligation for precision ... they can concentrate, in the text, on the second obligation, clarity.

Making the tables self-supporting means using full, descriptive titles and row and column headings that are informative. A caption for a table that just says 'Milk production of treated cows' is quite inadequate. The caption must always designate the number of the table, even if there is only one in the article, and it should give the essential details of what is in it. Thus, a title:

Table 4: Milk production in litres/day of Jersey cows during the first 30 days of lactation after injection of iodinated casein ...

is far more informative. Above all, it means that the reader does not have to seek information from the text to work out to what the numbers in the table are referring.

Row and column headings should also be complete. A table in which the descriptive headings have been replaced with indecipherable codes can't be considered self-supporting. Not only should these headings be readable, they should also specify the units, for example, g/day, ml/100 mg, or %. By this means, the body of the table does not become cluttered with units and shows numbers only. In fact, if all the units in all of the cells of a table are the same, this information can appear in the caption as was done with *litres/day* in the previous example. If there are any omissions or abnormalities, such as missing data or unusual circumstances that might affect certain of the data, they should be explained in footnotes. Incidentally, tables and figures are the only place in scientific articles where footnotes are acceptable. Footnotes should also be used to explain abbreviations, symbols, references, and statistical information, even if these explanations are also given in the text. A footnote that invites the reader to 'see text for details of treatments' implies that the reader is clever enough to read the text and the table simultaneously, which they cannot.

Good tables should present the numbers in a way that highlights the patterns, features and exceptions in the data.

A good plan for authors is to ask colleagues if they understand fully what their table is about without referring to the text. If not, they must make it self-supporting by adding material to either the caption, the headings, or the footnote.

Graphs or tables?

Graphs or figures are often thought to carry more impact than tables, especially where continuous changes associated with continuous inputs of treatments are being described. Even so, it is sometimes difficult to decide which is the more appropriate. On the face of it, graphs that are simple are usually easier to digest than a group of numbers but are far less precise. In broad terms, if you aim to use the material to show *qualitative* features of the data and gross differences, graphs are ideal. If the testing of your hypothesis calls for a close, *quantitative* analysis of the results, then a table containing the exact numbers is a better presentation. For example, if we wished to show that wool production of sheep increases with increasing concentration of protein in the diet up to 15% and is not stimulated further by concentrations higher than 15%, the story could be simply and completely told by a graph in which the horizontal axis shows the concentration of protein. Here, we would be more concerned with trends than with absolute numbers, so the scale on the axis is not critical. The purpose of the graph is to simplify and to make the data more visual. On the other hand, graphs are almost useless when a detailed numerical analysis is important. If we wished to demonstrate that the daily requirement for wool production in a sheep is 1.7 g nitrogen per g wool, the precise numbers from which this estimate was derived would be essential and a well laid out table would be the appropriate medium. Graphs are showy but they do not allow you to summarise the results as well as tables.

So, before you opt for a graph and dismiss tables as boring substitutes, consider the views of A.S.C. Ehrenberg who is not only a strong advocate for tables instead of graphs but has presented an approach to constructing them that supports his opinion. (Ehrenberg A.S.C, 1982, *A Primer in Data Reduction. An Introductory Statistics Textbook,* Wiley & Sons, Chichester.)

SCIENTIFIC WRITING = THINKING IN WORDS

According to Ehrenberg, tables need not be just rows and columns of numbers. Good tables should present the numbers in a way that highlights the patterns, features and exceptions in the data. By inserting marginal averages and placing them so that contrasts and comparisons are easy, you can make the reader aware of the major balance of the data at a glance. For example, if you decide that the discussion may involve the contrast between two particular means, those means should be positioned in a table so that they are close enough together to compare visually. If the object is to show the relationship between two or more series of numbers, arrange them in columns rather than rows. Reading down columns is much easier than reading across rows and the patterns emerge more quickly.

For example, compare the following two tables. They both show the same data—the sales of lemonade at seven retail outlets in the first five months of 2010. Table X lists the retailers logically in alphabetical order and faithfully presents the sale in thousands of litres to two decimal places, but it is far from 'user-friendly'. Table Y, on the other hand, sets out to make these same data visually comprehensible by using at least four helpful techniques.

Table X: Sales of lemonade (in '000 litres) by seven retailers in Drysville during each of the first five months of 2010

Retailer	Month				
	January	February	March	April	May
Family Store	16.54	19.38	19.88	16.59	21.62
Nancy's Bar	206.48	274.56	275.98	213.78	303.35
Pizza Shed	63.54	77.82	81.76	54.21	89.49
Railway Café	29.70	30.79	33.53	27.41	34.64
Royal Teahouse	142.63	137.6	171.79	162.40	194.26
Ted's Supermarket	137.63	129.17	149.38	117.21	183.40
The Oasis	47.32	51.83	53.73	49.10	60.23

Table Y: Sales of lemonade (in '000 litres) by seven retailers in Drysville during each of the first five months of 2010

Retailer	Month					Average
	January	February	March	April	May	
Nancy's Bar	206	275	276	214	303	257
Royal Teahouse	143	138	172	162	194	162
Ted's Supermarket	138	129	149	117	183	143
Pizza Shed	64	78	82	54	89	73
The Oasis	47	52	54	49	60	52
Railway Café	30	31	34	27	35	31
Family Store	17	19	20	17	22	19
Average	92	103	112	91	127	105

First, the author has estimated that the extra information in the two decimal places shown in Table X is completely inappropriate when describing differences in the sales of lemonade and has rounded the precision to a manageable, but appropriate level.

Second, the visual impact of the relative success of the seven sites in selling lemonade is emphasised by placing the sites in decreasing order of productivity. Maybe there wasn't so much logic after all in having the retailers in alphabetical order.

Third, a small but discernible gap has been added to emphasise the clear differences between the top three retailers, which were major sellers and the other four which sold far less.

Fourth, row and column averages have been inserted to orientate the reader better. Again, a visual gap has been left to distinguish the averages from the rows and columns of data. The relative success of the retailers in selling lemonade is confirmed by the row averages, and new information that illustrates the differences between months is added in the column averages. Note how quickly you can see the very high productivity in May. Ehrenberg emphasises the desirability of averages rather than totals at the ends of columns and rows because averages can be readily compared with values in the body of the table.

In short, Table Y has deliberately made obvious the patterns and exceptions that the data have to offer. The table is telling you the results almost before you read the figures. In effect, it could be described as a graph made up of numbers.

Just as data from tables should not be quoted verbatim in the text, histograms or graphs should not duplicate data already given in tables. You have to make up your mind which suits your purpose best. Your purpose is, of course, to tell your scientific story clearly and convincingly so this will be the basis on which to decide. Editors dislike wasting space and money on duplication and, even more importantly, readers are not happy about being forced to read the same thing twice only to work out that they are not getting anything new the second time around. Repetition should be reserved for oral presentations and for a very different reason, as we shall see later.

Use of statistics in presentation of results

Statistical analysis is a powerful tool that allows you to place probabilities on your results. It prevents you from getting carried away with differences that could be due to chance and it gives support to statements that claim that treatments are effective. But, remember that the responses or the differences are the important things, not the statistical technique that has given you the confidence to claim them. For example, a statement like, 'analysis of variance showed that there was an effect of treatment (means and sem, df, p)' suggests that you didn't know that there was a difference until the statistics told you so. If you changed it to 'treatments differed significantly' or, better, 'treatment A was significantly better than treatment B (means and sem, df, p)', you would be relegating statistics to their rightful place as a tool and emphasising the experimental result. You should ensure that levels of probability are clearly stated but you don't need to present tables to describe how you derived the levels of probability. They are no more essential to your paper than intermediary

SCIENTIFIC WRITING = THINKING IN WORDS

chemical analyses may be to your final conclusions about chemical constituents. But, as with chemical analyses, you need to have justified that they were the proper tests and were properly applied—usually in the *Materials and Methods*.

There are good techniques that enable you to present the statistical information in the same tables or graphs you are using for presenting the data. For clarity, it is preferable to summarise large masses of data by reducing them to totals or averages. Averages are preferable to totals because they present the summarised data in the same scale as the individual values and this makes visual comparison much easier. If you do this, you should also indicate the degree of variation in your original data by presenting the standard error of means or the standard deviation of individual records. These are not the same thing, so you should not simply write '12.6 ± 1.3' because it is not clear whether 1.3 is the standard error or standard deviation. Putting (SE) or (SD) in parenthesis behind the '12.6 ± 1.3' can clear this up. Even when numbers in tables are arranged to allow a clear visual comparison between them, an order of statistical significance of any differences should also be included. Without it, results cannot be reliably interpreted. The word 'significant' has other meanings than the statistical one but, because in modern scientific literature, 'significant' usually means 'statistically significant' it is a good idea to avoid using it in any other sense.

> Discussion *means the discussion of* your *results and not those of others.*

The *Discussion*

Here, at last, you have reached the part of the article where your thinking and your interpretation are put on display and you can give your readers a chance to assess your qualities as a scientist. Up to now, you certainly had to think but all the sections contained material that was strictly circumscribed and you have had, perhaps surprisingly, little scope to express your views or ideas.

What makes an effective *Discussion*?

The first thing to appreciate is that *Discussion* means the discussion of *your* results and not those of others. You discuss *your* results in relation to those of others and, possibly, in relation to the 'real world'—their applicability to some practical situation or to the wider sphere of your scientific discipline. Therefore, it is *not* a section in which you launch into a review of the literature on the subject. All literature that you cite must be there because it supports or adds meaning to arguments about *your* results. For example, a statement of the form:

'Brown (2005) found X, but Black (2006) found Y. I found Y so my results support those of Black.'

will not do. This is telling the world about the wrong sets of results—theirs, not yours. Rather you should be saying:

'I found Y and therefore my results are supported by those of Black (2006) but not by those of Brown (2005) who found X.' In other words, begin with what *you* found and continue with how other people's results matched yours.

You might consider it presumptuous or at least out of chronological sequence to say that Brown and Black, who probably did their work long before you did yours, are supporting or not supporting your findings. The fact is that this is not a chronological review of the literature but a discussion of *your* results and you are not only justified in expressing it in this way but you are obliged to do so to fulfil that principle.

The second principle that you are obliged to follow if your *Discussion* is to be part of a worthwhile scientific document, is that every argument you develop within it must end in a conclusion. The whole purpose of your *Discussion* is to draw conclusions. Readers of *Discussions* in scientific articles need to be satisfied with what they read. You need constantly to help them feel that they are finding out something worthwhile. If you don't, they lose interest, stop reading and move on to read someone else's paper. That means that you have to go well beyond just presenting them with a few, seemingly random observations or comparisons with what other people may have found. You have to help them by developing your points of discussion to draw conclusions from each issue that makes up your *Discussion*. The conclusion can take many forms. It may be a summary of the issue you are discussing that incorporates your new findings. It may be a recommendation. It may be a piece of speculation that could act as the base for a new hypothesis to test in future experiments. It may be a statement of a new principle. Or you may conclude that you do not yet have enough evidence to make a conclusion—in this case you must be specific about what evidence still needs to be found and, possibly, how to find it. A statement that simply says:

... in good scientific writing, every paragraph must have a conclusion.

'I found Y and, therefore, my results are supported by those of Black (2006) but not by those of Brown (2005)',

is not a conclusion. It is a statement of fact. In this case, the conclusion would have to address your view or, better, your reasoned argument about why the results in one experiment differed from those in another and what that might mean for future research, or for the 'real world'. Even a statement that says, 'We do not yet have enough information about X or Y to be able to draw a conclusion', is far more satisfying to readers than to be left wondering about an unassessed list of contrasting or concurring information.

Here is where the traditional format of a paragraph is such a powerful help to your writing. It is an immutable rule of grammar that every sentence must have a verb. If there is no verb, it is not a sentence. There is an analogous rule, though less well known, that, in good scientific writing, every paragraph must have a conclusion. If there is no conclusion, it is not a paragraph in a scientific discussion. You may not find this rule in text books of grammar

SCIENTIFIC WRITING = THINKING IN WORDS

because I just invented it, but if you follow it, you are on the way to writing good *Discussions* for two good reasons. First, you will stand a good chance of keeping the reader interested and second, you will discipline yourself to maintaining a focus for your writing. If a piece of information does not warrant your drawing a conclusion about it, it is a dependable sign that it is not worth discussing at all.

Many writers who think that their paragraph is getting too long, often decide to move to a new paragraph for no other reason than to make it look about the 'right' size. If they do this, the reader will be left stranded and interpret the break as meaning that the writer had nothing more to say on the subject that the paragraph originally set out to address. A decision on whether to break a long paragraph into two is easy to make. If you can identify two conclusions, then develop each of them in separate paragraphs. If there is only one conclusion, then stick with the one paragraph even if it appears to be too long. In reality, complex arguments of discussion that have the potential to lead to long paragraphs can often be explained better if they are handled in stages along the way, each stage with its own conclusion. Once this sequence has been identified, the original, large paragraph can be split up appropriately and new, smaller paragraphs can be constructed correctly and effectively, each with its own conclusion.

Some journals have recently inserted new sections, called *Conclusions* or *Implications,* after the conventional *Discussion.* I suspect that they have done this in frustrated response to so many authors who try to discuss their work without concluding anything. These journals that have an obligatory section for *Conclusions* pose a problem to good writers because good writers will already have the conclusions as an integral part of their discussion. The solution is to extract these conclusions and restate them in a list in the new section. This is certainly better than spoiling the *Discussion* by somehow denuding it of its most important elements and saving them until the end to include in a separate *Conclusions* section.

What is there to discuss?

While you were collecting, processing and tabulating your data, you will have formed a number of ideas that might be developed in the *Discussion.* Some of these ideas are associated with the way in which your data relate to the work and thinking of other researchers that have been published in the literature. Some other ideas may relate to the way your new information will affect the 'real world', either its practical application or its contribution to the generic thinking in your scientific discipline. These ideas need to be developed and related to your data and to the literature in a logical way. Many will probably perish in the process, and this is normal, but some will come through as important features that you must discuss. These developed ideas are the 'arguments' that you will use in the *Discussion* because you must argue and justify them in the face of what is already known of the subject. You will have to present their limitations as honestly as you can. The *Discussion* then becomes a collection of such arguments about the relevance, usefulness or limitations of your experiment and your results, and the possibilities they open for new research. Each of your arguments will be a separate piece of logical writing and will normally be the substance of a complete paragraph. The technique of developing arguments in a *Discussion* is identical to that of developing a good paragraph.

Giving impact to your scientific story

Every evening in most big cities in the world, subeditors of the morning daily newspapers gather to decide on the news stories of the past day that they might print in their next edition. They usually have many more stories than they have space, so they first decide on the relative importance of all of the stories that they could use. Those that they deem to have low priority are culled and the rest are ranked for their likely interest for the readers. Then, they decide how to present each item of news to demonstrate to their readers the relative importance they have placed on it. They have a number of means of doing this. For example, the most important news always comes first; it is the front page news. It always occupies more space than the lesser news. It will usually be announced in headlines with larger font. Those headlines may be colourfully or dramatically worded. It may be associated with eye-catching pictures or illustrations. Even the type may be in colour. As a result, readers are left in no doubt about what is the day's most important news, even before they read it.

The technique of developing arguments in a Discussion *is identical to that of developing a good paragraph.*

How does all of this relate to your writing your *Discussion*? There is a strong analogy between the news editors' daily task and yours when writing a scientific paper. Some of your discussion points are inevitably going to be more important than others. You need your readers to recognise this if they are going to appreciate your overall discussion fully. But, even more telling, is the chance it gives you to structure the discussion in a way that will ensure that even ephemeral readers who just read the first few lines of your *Discussion*, will grasp your main message.

If the key to a well-structured *Discussion* is to sort out which are the most important arguments, then you need to have a systematic means of assessing the priorities of the arguments you intend to use. As you begin assembling these arguments you will become aware that some are likely to have more impact than others. So, out with the notepad and pencil and set down in note form all the arguments that you expect you might use in the *Discussion*. Then, clarify for yourself the relative impact of each. Examine each one thoroughly and give it a grading. Four categories similar to those we talked about in the *Results* are probably sufficient.

Category 1. Those arguments that are relevant to the original hypothesis and which allow you to make a positive statement of acceptance or rejection.

Category 2. Those arguments that are relevant to the original hypothesis but which for some reason are equivocal, or which lead you to suggest further experimentation or observation before acceptance or rejection.

SCIENTIFIC WRITING = THINKING IN WORDS

Category 3. Those arguments based on your results, not relevant to your original hypothesis but which you consider sufficiently new or interesting to be worthwhile including.

Category 4. Those arguments based on your results, not relevant to your hypothesis and of marginal interest.

Once you are satisfied that you have your ideas in an order of importance, you can take further advantage of the techniques of the newspaper editors. Unfortunately many of the tools that they are able to use to emphasise the importance of one article over another, like larger font, eye-catching pictures and stunning headlines, are not available to the authors of scientific papers. Editors of scientific journals tend to be more conservative than editors of newspapers. But two of these tools are very appropriate; positioning the best arguments first and making sure that lesser arguments take up less physical space than major arguments. You should use them because they give the right visual impression of priority of arguments to the readers.

… position— putting the argument where the reader is likely to expect it to be.

The first move is to cross out all arguments in category 4 or any that you could not easily classify. Those that remain become the basis of your *Discussion* and you have classified them in descending order of importance. You must now make certain that you present them to readers so that they gain an impression of the relative order of importance that is as close as possible to your own. You are well on the way to doing this if you can give readers the impression, by visual impact alone, that the piece of information they are reading, or are about to read, is important even before they have absorbed the contents. There are two techniques that make this possible, and by using them you can stimulate the reader into ranking the priorities of your arguments subconsciously in much the same order as you have. The first technique relates to size. The reader automatically relates the length of text devoted to an argument to its importance. Newspaper editors use this technique on the front page, but they are able to increase the area and importance of a news item simply by using big type. You can't do this and, to make things worse, your most important point may be the most simple to develop. If it takes only a few lines, and a minor argument takes three-quarters of a page, your whole *Discussion* is visually, and probably logically, unbalanced. This does not mean that you should react by adding irrelevant sentences to your main item to increase its size relative to the minor argument. Rather, you should ensure that minor parts of the *Discussion* are dealt with in one or two sentences so that they do not get undue emphasis. Of course, it is unlikely that a strong argument will occupy only a few lines. It will probably have several implications and applications, each of which must be developed in the argument or perhaps in several arguments. If not, it is probably a sign that you should think twice about its ranking.

The second technique is position—putting the argument where the reader is likely to expect it to be: the most important first and the least important last. Some authors mistakenly think that they should save up their most powerful argument to make an impressive finish to their discussion and, in so doing, leave a good impression. The problem is that busy readers often don't read the article to its end and the chances of their not persisting to the end are greater if what they read in the early part is dull and unexciting. In fact, the reverse is true; if they find interesting stuff in the first paragraph of the *Discussion* they are more likely be encouraged to keep reading. Newspaper editors recognise this and never leave their main points until last and neither should you. Their most important point is made very clear in the first few lines of their article and yours should be the subject of the first paragraph of the *Discussion*. Newspaper articles seldom end in a flourish of exciting revelations as novels or short stories do. They seek to inform, not impress, and that is also the role of a good *Discussion*. Some authors also think it desirable, and possibly essential, to tie up as many loose ends as possible before coming to the main points. So, they dedicate precious space to possible weaknesses in their methodology or unusual happenings during the conduct of the experiment before presenting their main conclusions. From the readers' point of view—and, in writing, theirs is the only point of view to consider—there is nothing more frustrating than to be presented with a battery of trivia when searching for main conclusions. Unless they have unlimited time, or an unusual interest in the work, they will read your first paragraph and assume that your loose ends are all you have to discuss. They will turn to the next paper without bothering to get as far as your important material.

Despite the best of intentions, complying with the techniques of using size and position to convey impact vicariously, may not be always possible. If so, it does not hurt to state 'The most important issue arising from my results is …' Do not overdo this, because most readers quickly tire of hearing of how important you think your work is. It is a technique to use, occasionally, when you feel that you must offset the fact that the size and position of the argument in the *Discussion* may be inadequate to convey the emphasis and relative importance that you want.

The paragraph as a vehicle for your arguments

We have seen already that paragraphs in scientific writing must end in a conclusion. The paragraph is, then, the development of the argument towards that conclusion. Physically, it gives readers visual help, in breaking up the total bulk of the *Discussion* and allows them to absorb your points one at a time. When they reach the next physical break in your *Discussion*, they should be able to pause momentarily and be satisfied that they grasped the implications of the section they have just read. If your paragraphs are not properly assembled, they will not satisfy readers who will be frustrated by what they are attempting to read and will give up.

A good paragraph has three components, a topic sentence, a logical development and the conclusion.

The topic sentence. Reading is much easier and more effective when we have some idea of what we are about to read. So, we need to start the paragraph with the topic sentence which is a mini-summary of what is to follow. It may, in fact, paraphrase the main point you wish to make in the paragraph. It immediately attracts attention and puts the reader on the right mental wave-length to receive the ideas in the logical developments that follow.

As well as signalling the substance of what follows, the ideal topic sentence should also act, whenever appropriate, as a link with the previous paragraph if there is one. This enhances the coherence of the whole discussion. 'Fruit picked in the early summer not only lacks the colour of later-picked fruit but it is lower in soluble carbohydrates …' 'In contrast to the increase in heart rate of athletes, their appetite is unaffected by high temperature …' Both of these meet the two roles of referring to the previous arguments (dealing with colour of fruit, or heart rate) and signalling the subject matter of the new paragraph (which will discuss soluble carbohydrates, or appetite).

Many authors wonder whether it would be a good idea to put subheadings in the discussion to assist the reader. In fact, if the paragraphs have informative topic sentences, there is no need for subheadings because they play exactly the same role while, at the same time, allowing links and cohesion that are difficult to make with subheadings.

The logical development. The main body of the paragraph uses facts from your results and combines them with other facts or theories to make your point. Your aim is to draw a sound conclusion by deduction, induction or a mixture of both. For example, you may believe that your results allow you to make a generalisation not previously possible. This would be developed by assembling the essence of your results and, possibly other results from the literature, by the process of induction. You may also feel that your results have a certain new application and your argument, to demonstrate this, will be based on deduction from your own and others' results.

The conclusion. This is a statement of the message that you wish the reader to retain from the issue that you have been discussing in the paragraph.

For example:

'If fruit is to be sweet enough to process it must not be picked before mid-July.'

'There is thus no reason to believe that athletes will eat less when exposed to high summer temperatures.'

Here we have examples of a specific and a general, concluding sentence. Both of them have a clear message which they deliver emphatically. The reader will, of course, want to be convinced of the reasoning behind the conclusion. It ought to be there in the body of the paragraph if the argument has been well developed.

In practice, it is a good idea to try out the sequences of information in each argument in the form of notes. In this way, you can decide finally on what sequence seems most logical and can therefore be understood most quickly by a reader. Clearly, the key sentences are the first, which says what the paragraph is about and the last which is your main message for the reader. Once you have expressed these, which may take a lot of thought, you will be amazed at how simply you can write the sentences between them because you have a point of departure and a destination for your writing. You can easily judge whether each potential sentence is on or off the track towards the conclusion and quickly either write it with confidence or dismiss it as irrelevant.

Nevertheless, each argument is unique and is supported by its own set of information; so there can be no specific rules except that you must avoid some common fallacies of logic.

Generalisations: An invariable property of biological, sociological or medical data is their variability. Thus, premature generalisations based on a few preliminary results, or on results obtained under a limited set of conditions could lead you to make statements that might embarrass you later. A valuable use of statistical analyses is to minimise the chances of making foolish generalisations. If the early evidence points your way, you have reason to be enthusiastic but do not let it tempt you into rash conclusions that put your logic into question. Even when statistical analyses show that your results are unlikely to be due to chance, your generalisation must be guarded so as not to exceed expectations based on the limitations of your data and how and where they were gathered.

Authority: There is no ultimate authority in science. Even Newton's laws of conservation of energy, which stood for centuries, were modified by Einstein, and Einstein himself is occasionally being challenged. However, the cornerstones of most scientific arguments are one or more authoritative sources. It is impossible to go back to first principles in every case. So we have to acknowledge certain concepts and statements as being acceptable, for the time being, as authoritative sources. So be careful that your choice of authority does not jeopardise your arguments. If your source of authority is out of date, or controversial, or simply wrong, your whole argument crumbles. In modern science, principles and concepts are being constantly revised in the light of new evidence. Your very paper may be presenting such evidence, so you are obliged to get it as close to the scientific truth as you can. You must be sure that the principle on which you are relying is currently accepted and recognised. An article in a peer-reviewed journal comes closer than most other sources to this criterion. If you have reservations about the authority you are quoting but can find no better alternative, your reasoning should be appropriately modified to make this clear.

Expressions of confidence: Your conclusions should be expressed according to the force of your data. If you have no conclusive evidence don't dither around with expressions such as 'It may be possible that …' or (worse) 'The possibility exists that …', which immediately suggests that you do not believe your own data anyway. On the other hand, slight differences between treatment and control groups do not permit you to say 'There is a clear indication …' or 'There was a marked response …' In these cases it is safest if you do not develop your argument beyond giving the actual values. This is honest, factual, and eminently scientific.

The best way of getting a message across is to make sure that its meaning is clear the very first time it is read. The first step is to construct the paragraph so that its topic sentence signals precisely what is to follow. Following this, the sentences should be simple, readable and in logical sequence. At this point, consider carefully the principles of 'reader expectation' on page 64. Impact is lost if sentences are woolly, or flowery, or ambiguous. The criterion of clear writing should apply to every sentence throughout the paper of course, but it is especially vital here in the *Discussion* where you are striving to make your arguments stand out.

SCIENTIFIC WRITING = THINKING IN WORDS

Speculation in the *Discussion*

Speculation is one of the most controversial aspects of scientific writing. It is defined in the dictionary as 'a conclusion, theory or opinion based on incomplete information or evidence'. Editors disagree among themselves and with authors or referees about how much speculation is permissible. With some it is entirely forbidden. The word 'speculation' scrawled across a paragraph not only denotes its rejection but conveys the impression that the work is unworthy, with overtones of charlatanism. I cannot accept this view because I believe that good speculation is the spice of science. It is the means by which new untested ideas can reach a wide audience. Also, it often stimulates readers to generate ideas of their own or provide the missing information or evidence to convert a piece of speculation into a respected conclusion.

But let me qualify my stand here. When discussing the development of hypotheses, we saw that an hypothesis is an idea that fits the known information and which can be, but has not yet been, tested. It is, in fact, a form of speculation. The most fastidious of editors will accept it in the *Introduction* because it is to be tested and validated or rejected in the body of the paper almost immediately after it is proposed. Speculation in the *Discussion* is left untested and so invites criticism. I believe that, if a piece of speculation is developed from the results of the experiment in the same way as an hypothesis was developed in the *Introduction*, and meets the same criteria as an hypothesis, it is not only acceptable in the *Discussion*, but desirable. The only valid grounds for rejecting a piece of speculation are that it does not fit the known information, including that described in the paper, or that it could not be tested using known technology. These constraints are sufficient to prevent undisciplined theorising but, within them, authors have the opportunity to raise new ideas and new ideas are the rarest and most valuable of all scientific resources.

Editors of scientific journals agree that one of the most common faults of Discussions is that they are too long.

The length of the *Discussion*

Editors of scientific journals agree that one of the most common faults of *Discussions* is that they are too long. And that usually makes it tedious to follow. Common reasons for *Discussions* being tedious are unnecessary references and spurious sentences that don't lead to the conclusions being made. If the conclusion is kept clearly in mind when the body of each paragraph is being written, there should not be a problem.

Another source of excess material in the *Discussion* is data repeated verbatim from results. The *Discussion* is meant to be read in conjunction with the *Results* and so repetition is seldom necessary. You can always refer to tables and figures already presented in the *Results* without otherwise repeating them. In any case it is very difficult to read fluently a series of numbers given in the text. The *Discussion* builds on the information in the *Results* by generalising, making comparisons and drawing conclusions. Where exact values are essential, only one or two key numbers should be used to avoid cluttering the argument. If you find that you have trouble making sense without re-quoting large segments of the *Results,* it is a sure sign that your results are inadequately presented. In fact, most writers find that they need to modify the layout of the *Results* when they work out the final details of the *Discussion.* This to-and-fro adjustment is a normal part of your final editing. It may involve changing a table to a graph to emphasise a trend you want to highlight. It may mean altering or exchanging row and column headings to point up a contrast or a pattern between figures. It can also mean deleting a lot of unwanted data from a table so that the uncluttered remainder can be used directly in the *Discussion.*

Citations in the *Discussion*

References have a very important role in scientific writing and their use and citation should reflect this. Statements like 'There is general agreement that … ' or 'The literature suggests that …' without references will just not do. Every statement you make must be supported by your own results, the results of others, or an authoritative statement based on the results of others.

References may seem to be relatively unimportant and minor to the main flow of the text, but unless they are accurate the validity of the text can be ruined. That is, names must be spelled accurately, data must be correct, and citations in the text must correspond completely with those in the *Reference* section. They must also supply the correct information. It is not sufficient to use a reference to a paper that is 'near enough' to what you are talking about. Readers wishing to follow your arguments more thoroughly should be able to find exactly what they are looking for in the paper to which you have directed them. If the paper is not the one from which the original information came, but merely one that used the original information to develop other arguments you have tricked the reader who may mistrust your interpretation as a result. Even worse, you may have offended the creator of the original information. To avoid errors of this sort try to get photocopies or off-prints of all material that you reference so that you can check it as you go. References are then available instantaneously to check that, for example, when you say Jones did something, Jones did in fact do it and, moreover, recorded it in the exact article, and in the exact year that you have cited. Your memory can play tricks, and libraries, even when just down the corridor, are sometimes too far away to check information straightaway.

References have many uses. They can be used as the ultimate authority on which to base arguments. They can be temporary authorities whose validity you intend to challenge, or you may consider them to be obviously wrong. It is possible to suggest to the reader which of these uses you wish to make of a reference by the way you word the text. Examine these statements:

All aerobic bacteria are sensitive to umptomycin (Bloggs 2007).

The implication here is that this is an accepted concept. Bloggs was the first to present it, and you, the author, agree. This emphasis is characterised by the placement of the author's name and the date in parenthesis at the conclusion of the statement.

SCIENTIFIC WRITING = THINKING IN WORDS

Bloggs (2007) found (or showed) that all aerobic bacteria are sensitive to umptomycin.

This implies that this is a less well-known concept, Bloggs discovered it, and you agree with him.

Bloggs (2007) claimed that all aerobic bacteria were sensitive to umptomycin.

In this case you imply that Bloggs went against general opinion with this claim and you are, for the time being, retaining an open mind on the subject. Note how the word 'claimed' and the use of the past tense indicate doubt and open up the possibility of a change of idea in the light of more recent work. Subtle changes of emphasis such as these can establish clearly your position relative to that of the authorities you are quoting.

Checking the logic of the *Discussion*

It may take you some time to get yourself to a stage in the writing of your *Discussion* where it is beginning to take a shape that pleases you—maybe days, weeks or even a month or two if you are working part-time on your writing. In that time you will have done a lot more thinking about the whole paper and you will understand much more about its strengths and weaknesses than when you began. It is not surprising, then, that in many cases the details of the way that you structured your *Results* and your *Introduction* will be a distant memory. You may even have changed your approach and emphasis as your understanding developed during the process of thinking and writing. But, as you approach the end of your first draft, you are now in a position to view the whole of the article instead of a small section of it and the necessary adjustments become much more apparent. You can go back to the *Results* and read them side by side with your freshly developed *Discussion*. When you do, don't be surprised to find that you placed some material and emphasised it in a way that is not as uniform as you might like. Now, you can edit to ensure consistency and make certain that *Results* and *Discussion* reinforce one another. Similarly, you can check your *Introduction* to verify that the logic and substance of your original reasoning matches the logic and reasoning of the conclusions you make in your *Discussion*.

The *Summary* or *Abstract*

The *Summary*, also called in some journals the *Abstract* or the *Précis*, is a kind of mini-paper that distils your full paper into a fraction of its original space. The summary is what most people who have been attracted by your *Title* will read next to flesh out their expectations. Some will be attracted to expand on what the *Summary* says and read the rest of your paper in detail, many will read no further. So, it must not only be concise, it must be complete. The challenge to authors is that editors of some journals set strict limits to the size of the *Summary*, either by setting a maximum number of words, usually between 150 and 250, or a proportion of the size of the paper, for example around 5%. To comply with these limits, you have to make some hard decisions about what to include and what to leave out so that you end with a concise mini-paper. Fortunately, if you wait until you have finished your first draft, you have already made these decisions and have already prepared most of the text that describes them. As a consequence and contrary to what many people imagine, the *Summary* is one of the easiest parts of your article to write.

There are a few simple principles to consider.

First, the *Summary* is often extracted from the rest of the text and may be physically separated from it in abstracting journals or on-line material. Even when it is with the rest of the text, many readers may confine their reading of the article to just the *Summary*. It therefore has to stand alone and this means avoiding references because they cannot be checked in the bibliography. Similarly, acronyms or abbreviations may not make sense without the text of the full article.

... contrary to what many people imagine, the Summary *is one of the easiest parts of your article to write.*

Second, the *Summary* never appears anywhere without its *Title*. This means that information in a well-crafted *Title* can be used as part of the written *Summary* and save you precious words.

Third, to be effective, a good *Summary* should provide the reader with four distinct components. You must expect that many readers will not read beyond your *Summary* and you will be selling yourself short if you don't provide them with all four of them.

1. **Why** you did the experiment

2. **How** you did your experiment

3. Your **major finding** or findings

4. Your **major conclusion** or conclusions about those findings.

Constructing the *Summary*

What goes into each of these four components?

Why? If you copy the hypothesis that you tested and make it the first sentence of your *Summary*, you will present your motivation for doing the work in the most effective and economical terms. You don't have the space to reiterate any of the reasoning behind the hypothesis but the hypothesis itself is the conclusion of that reasoning and can be expressed usually in one single sentence.

How? This is a broad description of the experimental design you used to test the hypothesis.

We measured the acid balance and peptic index of primary school children in Snake Gully province over three successive years.

We analysed the rate of recovery from hyperventilation of patients at three temperatures following a standard Peabody stress-test.

SCIENTIFIC WRITING = THINKING IN WORDS

To conserve space, present only the methods and not the materials unless there is a special reason to do so, such as when a particular 'material' is the subject of the hypothesis. Often, writers will have already written the appropriate words for the 'How?', particularly if they stated the *aim* of their experiment, and how they intended to achieve it. These words may be transposed directly to the *Summary* to complete this section.

The major result or results. You will already have applied the appropriate tests to your data when, in writing your *Results,* you were deciding on priorities to determine which of them were in category 1. These, and only these, are the results to present in the *Summary*.

The major conclusion or conclusions. Similarly, you will have already decided on your most important points for your discussion. You can take the words of the last sentences of each of the paragraphs that you designated category 1 and use them verbatim or you might wish to edit them slightly.

Voilà. There is the draft of your *Summary*. And, nine times out of 10, it will already be within the constraints of space demanded by the editors because you have confined yourself meticulously to the fundamentals.

Several times in this book I have said that a well-reasoned hypothesis may be a challenge to elaborate but, once worked out, simplifies the rest of the writing of the article. Writing the *Summary* is an outstanding example of what I meant. With this groundwork done, it is possible, in most cases, to write an excellent and complete *Summary* in no more than 10 to 15 minutes. And, because over 80% of *Summaries* in published journals leave out at least one of the four essential components that *Summaries* should have, yours will be among the best.

With this groundwork done, it is possible, in most cases, to write an excellent and complete **Summary** *in no more than 10 to 15 minutes.*

The other bits

Authorship

Modern research is increasingly done by research teams rather than by solitary scientists and this means that the articles that result from the research usually carry the names of more than one author. The decision about authorship and arrangement of authors in such multi-authored papers is, unfortunately, too often a highly sensitive issue in writing the paper. Ambition, jealousy and egotism are sometimes involved of course, but because there are no rules, many of the misunderstandings are genuine differences of opinion and people are confused about how to come to decision. One solution that is sometimes advocated is to have all interested parties decide on authorship before the research is commenced or at least before writing begins. However, that often exacerbates the problem rather than solving it. One's final commitment is often quite different to one's proposed commitment, so consolidating the position of someone who finally contributes little can be a source of much discontent. In the end, there is no substitute for goodwill among collaborating researchers to resolve the issue of authorship. Regrettably, goodwill is not universal among researchers.

But, there are some principles that, if agreed to, can lead to sensible decisions about who should be authors and in what order those authors should appear.

The overriding principle is that a research paper in a learned journal is an intellectual assignment and the important processes culminating in publication are intellectual ones. That means that people who make a substantial intellectual input to the whole process should have an *a priori* right to authorship. In addition, if the relative intellectual contributions of the authors can be established, then this can be a reasonable and equitable basis for deciding the order of the authors—assuming of course that the first author is the senior author and the rest are in order of contribution.

In the end, there is no substitute for goodwill among collaborating researchers to resolve the issue of authorship.

What are the intellectual components?

- Idea or ideas that led to the research being carried out in the first place.

- Reasoning that converted these ideas into a testable hypothesis.

- Interpretation of the results in relation to the results of others and to the 'real world'.

- Drafting of the paper.

- Editing of the draft for reasoning and logic (rather than simply style).

- Fielding of comments from reviewers and preparation of the final version.

These six components are not all of equal importance, and will vary according to the field of research, so a group of authors, or potential authors, might decide to weight them differently.

You will note that this list omits several other components such as:

- Successful seeking of a research grant to enable the work to be done.

- Management of the department in which the work was done.

- Physical work in getting the results.

- Use of standard statistical or analytical techniques or the lending of special apparatus.

- Supervision of a student who did the work.

The components of this second list are certainly vital to the successful completion of a piece of research but, in themselves, are not part of the intellectual input. On the other hand, they may have elements of the first list within them. For example, it is difficult to imagine that an

SCIENTIFIC WRITING = THINKING IN WORDS

application for a research grant would be successful without a good idea or hypothesis behind it. However, if the idea was not that of the person who went through the routine process of applying for the grant, it does not seem fair that that person should claim to have made an intellectual input. Similarly, supervisors usually have an influence on the thinking and reasoning of their research students—good ones always do—but they should not be able to claim authorship automatically unless they specifically influenced the research being published. Heads of departments and administrators gain their kudos in other ways and ought not have the mandatory right to interpolate their names onto articles to enhance that kudos.

Once the six components in the first list are agreed and, possibly, some weighting given to each depending on the research to be reported, potential authors can submit a suggestion of their percentage input to each component and be allocated an overall score. In practice, the total comes to much more than 100% but, once this has been rationalised, the overall scores can be used to decide the final authorship more or less objectively.

However, this is only the second best solution—remember that there is no substitute for goodwill among collaborating researchers.

Acknowledgements

Most journals allow a section at the end of the text of scientific articles for the authors to recognise people or institutions that have helped the research project to a successful conclusion. It is usually the place to thank people who have made a physical rather than an intellectual input. A substantial intellectual input, as we saw above, is more appropriately recognised by co-authorship. People who worked on the project as part of their daily job and who did no more than that, should not expect to be acknowledged. However, at the other end of the scale, if technicians become involved in the research by doing much more than is expected of them they deserve recognition. If it means that they did more than the job entailed physically, for example, working long hours or weekends, then it is both courteous and just to cite and thank them. If it means that they helped to explain the results they got or participate in a new and enlightening form of analysis, then they ought to be considered for authorship.

Funding bodies love to see their names associated with successful research and, in the interests of your long-term survival, it is probably a good thing to indulge them here in the *Acknowledgements*. Nonetheless, it is not a place for random name dropping. If you are going to cite someone here, let them know in advance and get their permission. If you don't, and they would rather not be associated with your article, it could be very embarrassing. In any case and contrary to what some people may think, mentioning a luminary in the field as part of the *Acknowledgements* is unlikely to influence the comments of a reviewer or editor. It is the article itself that will do that.

The *Bibliography*

References are the essential support for logical scientific thinking and reporting. Solid peer-reviewed articles are the most acceptable starting point for arguments leading to hypotheses and conclusions because they have been through the process of scrutiny and approval by practising scientists in the field. Part of this process is to record in the text of your article every reference you used and report its origin correctly in the *Bibliography* section at the end of the article. This means that you are not obliged to justify the validity of a reference in same the way that you must do with your own, new data. However, there are some cases and some areas of research in which peer-reviewed references are simply not available. So less convincing and scientifically robust forms of reference such as newspaper articles, anecdotal information or common practice are the only information available. If you are forced to use these, then you should be careful to acknowledge their less rigorous basis and modify accordingly the force of the conclusions that you base on them.

References in scientific articles have a central role in scientific thinking and writing so it would be reasonable to imagine that there would be a consistent and logical form for presenting them. Unfortunately there is not. There are hundreds of permutations and each journal has its own rigid format which may be subtly or radically different from that of the next journal.

In general terms, there are two main methods of referencing articles in journal and book publications: the *Harvard* (author–date) and the *Vancouver* (author–number) reference systems. In the Harvard system, the author's surname and year of publication are cited in the text, for example (Scmurch, 2007), and all the citations are listed at the end of the article in alphabetical order by author. In the Vancouver system, citations are indicated by a series of numbers in parentheses. For example, *the moon is blue at the equinox (1), but Bloggs (2) found traces of red*. In the *Bibliography*, these are listed in numerical order as they appear in the text. Claims are made in favour of both systems, of course. For the *Harvard* system, the advantage is that a reader wishing to know the authorship of a piece of information is not obliged to decipher it from numbers. By contrast, it is claimed that, with the *Vancouver* system, the main text reads more easily, and some editors consider it to be less obtrusive. The *Harvard* system is more universally used, but the *Vancouver* system is firmly entrenched in many journals in medical science. Making the problem more complex, many publications often have their own house style that introduces specific modifications of these general standards and these are rigidly enforced. Therefore, writers cannot afford to ignore the *Instructions to Authors* that each journal publishes from time to time and must follow these meticulously when composing the *Bibliography*.

This book, which is attempting to concentrate on logical thinking and presentation in scientific writing, is out of its depth in the face of such chaos and inconsistency. But, fortunately, computers have come to the rescue. They can reproduce the requirements of almost any journal and alter a standard database of references to comply fully with the journal's idiosyncratic requirements. Software programs such as *EndNote*, *Reference Manager* and *ProCite* give writers the capacity to concentrate on the structure and logic of their articles and leave the computer to handle the tedious but very necessary responsibility of keeping the references organised. It is probably best to call on these programs for help at the end after you have identified all the references you plan to use and where they will be cited in the text.

Editing for readability and style

At this stage of the writing process your article should have its basic structure. It will go together strongly scientifically but it is likely to have some imperfections that need to be edited to improve its style. In other words, you have organised what you want your scientific story to tell, and you can now concentrate on editing it to ensure that it is fluent and easy for your fellow scientists to read. In essence, editing means that the passage of your thoughts and information from the page on which you have written them to the consciousness of the reader is not obstructed by the way in which you express them. Even if you are not a native English speaker, and even if you have written some parts of your draft so far in your own language, this is still not the time to call in a native speaker to rewrite the text for you. There is a lot that you can and must do before you relinquish control of the writing process to other people.

The first objective, of course, is to ensure that what you think you have said is the same as what the reader thinks you have said. That is why precision, clarity and brevity are critical qualities of all scientific writing and these three elements should already be part of your draft. These alone, however, may not be sufficient when writing for busy people who have a limited time in which to grasp your message as clearly you would like. They certainly don't want to spend time admiring your erudite turns of phrase or checking their dictionary for unfamiliar words or re-reading badly constructed passages to be sure that they comprehend them. Readers of scientific literature expect to understand and, you hope, be influenced at their first pass—not to indulge in an exercise in deciphering. When they want to do that, they take up solving cryptic crosswords or Sudoku puzzles. The more time they need to be certain of understanding the exact meaning of what is written, the greater the chance that they will misinterpret your meaning— simply because they are unlikely to take that extra time.

The first objective, of course, is to ensure that what you think you have said is the same as what the reader thinks you have said.

There are two major but reasonably simple ways that you can intentionally ensure that your readers find your writing to be fluent and readable. The first is to eliminate from your writing all those expressions and structures that create verbal stumbling blocks—points where readers are likely to abandon pursuing your thoughts to sort out the configuration of your words. The second is to understand sufficiently the way in which readers read so that you can deliver the words in a way that matches their reading and allows them to pick up your information as efficiently as possible. Neither of these two ways is difficult to use and they ought to be used routinely in your writing. Indeed, authors who consistently write well probably use these or similar techniques regularly without being consciously aware that they are doing so. Now, by being aware of them, those of us without such intuitive skills can improve the readability and fluency of our writing spectacularly.

Eliminating verbal stumbling blocks

If we agree to keep to the dictum that we are writing to inform and not to impress, we will automatically avoid some of the stumbling blocks that frequently characterise scientific writing. That simply means steering clear of words and expressions in our writing that we would not normally use if we were explaining the same thing verbally to a colleague. Despite this, there often remain cumbersome ways of expressing things that may seem familiar or okay when we write them and may even sound as if they are eminently scientific. However, other scientists in other institutions or other countries may find them unfamiliar, troublesome or ambiguous and that is an excellent reason to identify them, remove them and replace them with simpler and more easily understood alternatives.

Books on syntax and the use of words are full of examples of such expressions. When you read of the hundreds of expressions and grammatical pitfalls that you must avoid to write clearly you can be excused for being intimidated. But there are seven major ones and I am highlighting here what I believe are the most common stumbling blocks in scientific writing, together with suggestions for more acceptable options. Coincidentally, almost all of these are as relevant in other languages as they are in English. If you become familiar with just these few and deal with them whenever you are tempted to use them, you can guarantee that your writing will be clearer and easier to read. Of the hundreds that are left, most are 'fine points' that are not stumbling blocks that will distract most readers. For example, if you said that *less people responded to the treatment,* instead of, ***fewer*** *people responded to the treatment* you would be technically wrong but unlikely to be confusing. In other words, it would not be fatal when measured by its effect on comprehension and could be tidied up in the final polishing.

The seven verbal stumbling blocks

1. Clusters of nouns

Examples:

> *Soil nitrogen uptake.*
> *Annoying infant pathology problems.*
> *Starch absorption rate analyses.*
> *Artificial learning enhancement programs.*
> *Plasma urea nitrogen concentrations.*

These expressions are always cumbersome to read and fit precisely the dictionary's definition of jargon: '... language that is used by a particular group, profession or culture, especially when the words and phrases are not understood or used by other people'. Sometimes they are used in the belief that valuable space is saved by eliminating prepositions such as *of, on, in, for* and others. Sometimes, writers familiar with a certain group of nouns recognise them as one single entity and do not realise that a reader trying to assimilate them for the first time will struggle. Omitting prepositions may be permissible where the missing word is clearly understood. But, does *soil nitrogen uptake* mean *nitrogen taken up **from** the soil* or

SCIENTIFIC WRITING = THINKING IN WORDS

*nitrogen taken up **by** the soil?* These two alternatives are almost antitheses of each other. So what of precision and clarity?

Where several nouns are clustered and there is also a real adjective in the cluster, it is often hard to know to which noun the adjective refers. For instance, in the second example, are we dealing with pathological symptoms in annoying infants, annoying symptoms in the pathology of infants or symptoms with annoying pathology in infants? In other words, there are three quite different interpretations.

Fixing the problem:

There are three options here:

First, and probably easiest, is to replace the missing prepositions that have been assumed in the original draft. Prepositions, such as *of, by, in* and *from*, are among the shortest words in the English language. Inserting an extra one or two will not lengthen your article very much but will do wonders for its clarity. In fact, in the examples of noun clusters above that were ambiguous, we saw that the alternative interpretations were very different from each other but, after the insertion of one or two prepositions, each alternative was precise, clear and unique. In a list of the most commonly used words in the English language, no less than six of the top 20 are prepositions that you might use to fix noun clusters: *of, to, in, for, on* and *with*. It is hardly a radical technique.

Second, if there is an appropriate adjective with the same stem as one or more of the nouns in the cluster, replace the noun with the adjective. For example, *pathological symptoms*, not *pathology symptoms*.

Third, where words seem particularly appropriate together use a hyphen to indicate that they should be read as one composite noun. For example, *fine wool sheep* could mean either *sheep with fine wool* or *fine sheep with wool*— writing *fine-wool sheep* removes the ambiguity simply and completely. Another option would be *fine wool-sheep*, meaning fine sheep of a breed that produces wool. In each case all ambiguity has been removed. Free range eggs are certainly not gratis because they come from the range.

2. Complex adjectival phrases

Examples:

> *The maximum net returns above chemical treatment cost strategies.*
> *The high motivation but minimum social responsibility group.*

These two examples probably made instant sense to their creators when they were written and are most likely the result of an over-familiarity with the field. Unfortunately, once they are written, most of the people who then read them will be doing so for the first time and will stumble over them while they try to work out their meaning.

Fixing the problem:

As with noun clusters, these complex phrases can be resolved simply by using the extra words—mainly prepositions—to clarify their meaning. So, *The maximum net returns above chemical treatment cost strategies* becomes *The strategy that gives the maximum net returns above the cost of chemical treatment.*

Similarly, *The high motivation but minimum social responsibility group* becomes *the group that was highly motivated but showed little social responsibility.* In each case, clarity and precision have taken precedence over brevity.

3. Sentences beginning with subordinate clauses

Examples:

Although the results so far are for only a single ethnic group and the numbers are relatively small, laryngitis appears to be a consequence of too much talking.

Notwithstanding the fact that the spring in 2007 was warmer than average, which probably hastened the germination of seed after sowing, the physical size of the seed was strongly related to the rate of emergence of individual plants from the seed-bed.

In each of these cases, there are two subordinate clauses before the main clause.

Researchers are, by nature, cautious people. They are reluctant to make bold, unqualified statements because they know that their peers will scrutinise these carefully and expose their lack of precision or undue generalisation. So, they feel more comfortable presenting conclusions only *after* they have confessed to any possible qualifications to those conclusions—never *before*. This, of course, is commendable but readers, unlike the writer, have the simple problem that they do not know what the qualifying statement is about until they read the main clause at the end of the sentence. The longer the subordinate clause the worse the problem. In effect, readers respond to this by mentally ignoring the qualifying or subordinate clause while they search for the main clause. Then, armed with the key information in the main clause, they re-read the sentence to get its full meaning. Sure, they get there in the end but they have to stumble over the sentence twice to do so. Their short-term memory just can't handle it in one pass. The most important part of any sentence for guiding the reader is the beginning and that is where the key message should be placed.

Fixing the problem:

This is just a matter of suppressing false scientific reserve in the interest of readability and putting the main clause at the beginning of the sentence.

So, *Although the results so far are for only a single ethnic group and the numbers are relatively small, laryngitis appears to be a consequence of too much talking*, becomes, *Laryngitis appears to be a consequence of too much talking although the results so far are for only a single ethnic group and the numbers are relatively small.*

Occasionally, a condition or reservation may be the key issue in a sentence. In this case you are justified in placing the conditional clause first. For example, after a statement about the value of fertilisers, a writer may say: *If there is insufficient rainfall, it is uneconomical to apply supplementary fertilisers.* In addition, if the subject of the subordinate clause had been dealt

SCIENTIFIC WRITING = THINKING IN WORDS

with at the end of the previous sentence, then a subordinate clause at the beginning would not be a mystery and would be acceptable (as this sentence illustrates!). Nevertheless, when you put a subordinate clause first, be sure you do so for the right reason.

4. Nouns instead of the verbs from which they are derived

The action of any sentence is in its verb and every sentence must have one. It therefore makes sense, when attempting to write positively and clearly, to concentrate particularly on the verb and make it work hard for you. Despite this, many scientists, when they have a choice, prefer to use a noun rather than a related verb to make their major point.

Examples:

Weights [noun] *of the children were taken.*

Reductions [noun] *in temperature of the guinea pigs were seen 10 minutes after they were subject to immersion* [noun] *in iced water.*

These sentences may not be serious stumbling blocks but they are awkward and can invariably be made better.

Fixing the problem:

Look at each noun in the sentence and see if it has a verb derivative. If so, simply use the verb to replace the noun.

Weights [noun] *of the children were taken* becomes *The children were weighed* [verb].

Reductions in temperature of the guinea pigs were seen 10 minutes after they were subject to immersion in iced water, becomes *The temperature of the guinea pigs was reduced 10 minutes after they were immersed in iced water.*

You will notice that by transforming a noun into a verb from the same stem we have done two valuable things. First, we have automatically dispensed with the original verb, which indicates that it didn't have much value in the first place. In fact, it was only there to make the phrase into a sentence. Instead of *taken*, we could have used verbs like *noted, recorded, observed* or many others without substantially altering the meaning of the sentence. By contrast, in the new sentence, there are no other verbs that can satisfactorily substitute for the new verbs, *weighed, reduced* or *immersed*. Second, we have shortened the sentences.

And, because of these two improvements, we have sharpened the impact of the sentences. The remarkable feature of this easy technique is that it works 100% of the time. Replacing nouns with verbs is one of the most simple and yet most powerful tools you can use to improve the directness, clarity and brevity of your writing. The illustrations above were short and simple but look at this next, much longer and cumbersome sentence in which are underlined four candidate nouns that might be changed to verbs.

Increases *in ambient temperature resulted in a **deterioration** of the community's health status, particularly in regions where the **treatment** of the effluent ponds had not been carried out until the **commencement** of spring.*

When the four changes are made, it becomes:

When ambient temperature increased, the community's health status deteriorated, particularly in regions where the effluent ponds had not been treated until spring had commenced.

That simple, mechanistic treatment reduced a sentence of 33 words to one of 24 without losing any part of its information or even its manner of presentation. It could probably be improved further in other ways but changing nouns to verbs was a giant, first step.

5. Use of imprecise words

Precision is central to scientific writing, so imprecise words have no place. The problem with them is that the same word may be interpreted by different readers in an entirely different way.

Examples:

considerable, quite, the vast majority, a great deal, rather, somewhat, etc. and *and so forth.*

Considerable could mean anything from a few per cent to 99%. For example, if 100 people underwent a new surgical procedure and 10 of them died, that would certainly be a considerable number. If, on the other hand, a weedy field was treated with a herbicide and 10%, of the weeds died this would hardly conjure up the adjective, *considerable.*

Expressions like *and so forth* and *etc.* are often used when the writer cannot think of anything more to complete a sequence of words. This is the very antithesis of scientific precision. For example: *The data were treated statistically to take account of changes in temperature, humidity, daylength, etc.* Can you guess what *etc.* means here?

Fixing the problem:

It is invariably more specific and more helpful to give the exact figure or a rounded version of it. Thus, instead of, *A considerable number of plants responded* we should use *Seventy-four per cent of plants responded*, or even, *about three-quarters of the plants responded*. Remember, of course, that the rounded version would only be appropriate if the precise figure were given elsewhere, such as in an accompanying table or figure.

Avoid words like *and so forth* or *etc.* as a matter of course. If you insist, then the only time that you should use them is when the identity of the *etc* or *and so forth* is absolutely clear. For example, *The 20 aliquots were labelled 1,2,3,4, etc.* is unambiguous.

6. Use of acronyms, unfamiliar abbreviations and symbols

Acronyms are an increasingly prevalent and contagious infection of modern scientific writing. Many authors seem to delight in inventing new ones and using them wherever they can so that articles with as many as 20 or 30 different acronyms are appearing more and more frequently. Some journals have recently introduced a new section at the beginning of the article in which acronyms are to be listed and defined in alphabetical order. This is treating the problem without attacking either the cause or the consequences.

Whether in a list or not, it is hard to think of a more potent stumbling block to readability than an acronym. Even acronyms that are relatively familiar require a moment or more of distracting thought before the reader can continue on with confidence. Acronyms that are not so familiar, despite having been explained earlier in the article—or in a list—have a high chance of completely blocking the flow of information to readers who are obliged to spend time tracking down their meaning.

Despite all this, abbreviations can sometimes be useful. If an expression that could be abbreviated is to be used many times in a paper it should certainly be abbreviated. Even so, it should be written in full in the *Title*, in the *Summary* and in headings to graphs or tables—in short, anywhere that it might be read separately from the text in which it is defined. This also allows the reader more opportunity to assimilate it. If an expression is not used more than three or four times, the saving in space through abbreviation will in no way compensate for the time and concentration lost by readers while they verify the meaning of the abbreviation.

Commonly accepted and well-known abbreviations—which often may not be as commonly accepted or as well known as you imagine—are usually difficult enough for most readers. For example, AA means *amino acid* to biochemists and *atomic absorption* to physicists but it is also familiar as *Automobile Association* to motorists, and *Alcoholics Anonymous* to others (presumably not motorists!). On the other hand, new abbreviations that you invent yourself and present to the world for the first time may make you feel like some sort of pioneer but invariably disrupt readability and, because of this, should be avoided except in the most extreme circumstances.

Fixing the problem:

In short, whenever possible be frugal in your use of abbreviations and be aware of their catastrophic impact on readability. If you were a plant geneticist you might understand:

Location of BAC clones, together with CM localised by FISH and PRINS were combined with CACM to construct an idiogram of NLL.

If not, you would need several minutes at least to begin to comprehend what was being said. Readers don't usually have that sort of spare time and will most likely say to themselves that they don't belong to the club for which this is some sort of secret code and will stop reading altogether.

7. Citations, footnotes, asides in parentheses and other distractions

Citations, in particular, are essential for justifying much of the logic in scientific writing. But they have a cost in readability unless they are handled carefully. Don't allow them to fragment sentences unless you have a special reason. Consider this sentence:

The number of stomates per leaf may increase in geraniums (Brown, 1937), decrease in petunias (Black, 1978) or remain constant in sweet peas (White, 1990) when manganese is deficient.

This construction makes sure that each fact is accorded its appropriate author but the sentence is difficult to read because the cited authors have intervened too much. A more acceptable statement, because it is more fluent, might be:

When manganese is deficient, the number of stomates per leaf may increase in geraniums, decrease in petunias, or remain constant in sweet peas (Brown 1937; Black 1978; White 1980).

This bundles together the necessary references at the end of the sentence where the reader has the option either to look at the citations or to absorb the main message for which they have been quoted and move on.

Anything in parenthesis is a potential distraction. It says to the reader, 'This is less important than the main text, but I want to redirect your attention to it anyway because you might find it interesting.' Each time that you are tempted to put any information in parentheses, consider whether your story needs this information. If it doesn't, leave out the whole thought and let the reader get on with it. If it does, then you have all the reason necessary to incorporate it into the main text and eliminate the stumbling block that it may cause if you were to leave it in parenthesis.

Footnotes are in the same category as parentheses, except they are further from the text they are intended to support and therefore even more likely to dislocate its flow. Fortunately, most journals in the biological and medical sciences discourage their use but, in some social science publications, they are still alive and well. It is naïve to believe that by tucking away supplementary information in a footnote it will enhance the flow of the main text. Readers faced with half a page of text that is apparently the main story and another half that is footnotes have to choose between whether they follow text and ignore the footnotes or divert slavishly to the footnotes and hope to remember the train of thought in the text—or give up altogether. Giving up is an attractive option unless they have a very special reason to pick their way through the mess.

Fixing the problem:

The important thing is simply to recognise that diverting the reader's attention from one thought to another or one part of the text to another and back again is chaotic for readability. So, to minimise the chaos, avoid brackets, footnotes and appendices whenever you can. Either leave out the information if it is only incidental to your main message or, if it is an integral part of the message, reconstruct the passage to incorporate it into the text. In the case of citations, which custom and the *Harvard* citation system dictate must have the year or both the author and year in parentheses, try to position these breaks where they will be likely to do the least harm.

With a small amount of effort you can learn to recognise and deal with these seven major stumbling blocks as a matter of routine. There are, of course, many other faults and most people that you talk with will have a favourite that they like to treat as anathema. However, if you learn to take care of these seven and convert to acceptable alternatives, you can be confident that you have covered most of the blemishes that cause readers to falter when reading scientific articles.

An eighth stumbling block?—sentences that are too long

Many people regard long sentences as being difficult to follow and therefore potential stumbling blocks. Certainly, programs that check grammar flag them automatically as problems. But I don't think this is always the case and so I have not included them with the other seven. Some long sentences when carefully and logically constructed, especially using the concept of reader expectation on page 64, can flow well and convey their meaning easily in one pass of the reader's eye. By contrast, a series of sentences that are too short can often be annoying because of their staccato style. A combination of longer and shorter sentences is usually much more pleasing but, nevertheless, beware of long and complex sentences and check that they are indeed fluent.

Why are our written sentences sometimes too long? When we speak, we form sentences without even thinking about them and we rarely bother about the punctuation. Punctuation is reserved for sentences that we write. But if we want to make absolutely certain that our listener is following what we have to say, we pause frequently and use shorter sentences. On the other hand, when we write, that is not necessarily the case because our sole listener is ourselves. So, we adopt a different strategy because we don't want to lose the thread of the idea that we are developing and are continuing to develop until the end which often results in long sentences, especially in the first draft of the article. The reader, on the other hand, is not so immersed in the text as the author and has problems reading the whole sentence fluently. You probably found that the second last sentence you just read was a bit long. It's OK, I wrote it like that deliberately.

Fixing the problem:

A sentence is too long or too complicated generally because it has a main clause and one or more subordinate clauses. So, look for the conjunctions like, *and*, or relative pronouns like, *who* or *which*, and replace them with a full stop. Now, make sure new subject to the sentence is not likely to confuse the reader. Then, check that there is a clear link with the previous sentence as we see below and the problem is dealt with.

As an example, take the long sentence from the paragraph above. The possible points where it can be broken are highlighted.

*So, we adopt a different strategy **because** we don't want to lose the thread of the idea that we are developing **and** are continuing to develop until the end **which** often results in long sentences, especially in the first draft of the article.*

It then becomes:

So, we adopt a different strategy. We don't want to lose the thread of the idea that we are developing and are continuing to develop until the end. Unfortunately, this often results in long sentences, especially in the first draft of the article.

Delivering the written word in a way that matches the way a reader reads

A scientific article that presents all of the data and all of the scientific discourse that the author intended to present is not necessarily a successful article. It only becomes one when most of the people who read it can perceive accurately and quickly what the author really meant. For this to happen efficiently, the author has to be aware of what makes things easy to read. Gopen and Swan in *American Scientist in 1990* (Volume 78, 550–558) provided a brilliant insight into how this works; 'If the reader is to grasp what the writer means, the writer must understand what the reader needs.' They proposed the concept of 'reader expectation' which makes use of relatively new knowledge on how the reader perceives and interprets written information.

Basically, all information that we receive by the written word is either new or old. That is, it provides us with fresh concepts and ideas or else it consolidates ideas that we have already received. In most cases, we can find both types of information in the same sentence. The key to rapid comprehension is to use the old information to let readers know where they are in relation to what they have just been reading, and then, and only then, present the new information.

New thoughts are grasped much more readily when they are perceived from the comfort of what is already understood. So, the first part of the sentence should usually be used to make readers ready by linking them to previous information before the rest of the sentence discloses the new idea. The order is most important and we can systematically make great changes in the readability and the clarity of passages simply by getting the order right. If, at the same time, we take care to provide linking words that signal the substance of our next idea, we can almost work miracles with text that was previously tedious to follow.

You may already have recognised in this book the notion of generating within the reader an expectation against which he or she can compare new information. The concept works at the level of the whole article, where the hypothesis provides the expectation that can then be compared against virtually all the information that follows. It works, too, at the level of the paragraph where the topic sentence allows the reader to anticipate what is to be discussed in the rest of the paragraph. Now, it is the turn of the sentence where reader expectation is equally as effective at keeping the reader on track.

There are two ways of linking the leading words of a sentence to the older information that the reader has already taken in. The first is to repeat words that have already been used in the previous sentence, or, at least no more than two sentences before—or as Gopen and Swan put it, 'old' information. The second is to use linking words.

1. Repeating 'old' information

Here is a set of two sentences in which the second begins with a piece of information that has nothing to do with the first.

*The students were randomly selected and allocated to three treatment groups. **A new piffometer with twice the speed of old instruments** was used to monitor the speed at which students in the three groups learned to farnarkle.*

The information about the new piffometer is new information and might as well have come from outer space as far as readers are concerned until they read the rest of the sentence. Only when they have read the sentence can they work out where to put it in their train of thought. In practice, most readers would feel that they should re-read the sentence to be sure that they have placed the new information correctly in its newly found context.

If, on the other hand, the information in the second sentence were reversed and the 'old' material about the students presented first, readers could read the information more logically and comfortably in one pass and the wonders of the piffometer would be instantly apparent.

The students were randomly selected and allocated to three treatment groups. **These three groups** *were monitored for their speed of learning to farnarkle using a new piffometer with twice the resolution of old instruments.*

2. Using linking words

There are many words with which a sentence can start that instantly indicate the direction that the sentence is going to take. By using these words we can effectively connect back to the old information in the same way as repeating words from previous sentences. So, if a sentence begins with, *So …*, it implies that what is to follow is going to be a conclusion based on what has already been said. If it begins with *By contrast …*, it means that the rest of the sentence will be the opposite of what has just been said. Connecting words like these are like signposts, directing readers along the path the author wishes to take them and allowing them to plan in their minds to take in and retain detailed information efficiently, even before it arrives. Connecting words like, *Moreover, Notwithstanding, In addition, However, But, Therefore, Specifically, In summary,* and many others act as signposts in their appropriate place and they always improve the readability of what you write. For the same reason, if you have two or more possible explanations for, or consequences of, a particular result, tell your readers from the beginning so that they are well orientated with a statement like:

Using Gopen and Swan's principle of reader expectation is so simple that it is hard to believe it could be so effective.

There are three possible explanations for this result.

The simple, short sentence saying so is as good as a map that keeps the reader orientated through a relatively complex passage of information and keeps each piece of that information in perspective. Each piece will begin with, *First, … Second, … Third, …* and even if one or more of them takes more than one sentence, readers know exactly where they are and can follow the reasoning more easily.

Using Gopen and Swan's principle of reader expectation is so simple that it is hard to believe it could be so effective. But, it is most certainly easy to use and the results are invariably remarkable. The strategy is to develop a systematic approach to applying it. You will, with practice, automatically get the old or signalling information into the opening part of most of your sentences as you write them but, in concentrating on scientific content and logic, you may not always do so at the drafting stage. In fact, at the drafting stage it may be distracting to strive to do so and should not be your main priority. But, at the editing stage, it is almost a mechanistic process that you can do without much attention to the scientific content. Personally, I enjoy this procedure immensely because, when I am finished and I re-read the passage, concentrating once again on the science, I am often amazed at how much it has automatically made the writing more fluent.

Consider the following paragraph which describes short- and long-term memory.

Memory can be divided into two phases: short-term memory and long-term memory. When an animal learns something this information first of all enters the short-term memory where it will remain for a matter of minutes to hours. The experimental methods used and the species of animal studied can affect the precise duration of short-term memory. A number of agents including electro-convulsive shock (strong electric shocks applied to the head), low temperature, coma and deep anaesthesia can disrupt information that is being stored in short-term memory. Any of these treatments may produce a state known as retrograde amnesia, in which the memory of recent events is disrupted leaving earlier events unaffected. Since more remote memories are resistant to disruption, it has been concluded that the mechanism by which the information is stored in short-term memory differs from that for long-term memory. Because short-term memory is disrupted relatively easily by procedures which may be expected to have a profound effect on the electrical activity of the brain, it has been suggested that information is stored in short-term memory as reverberating electrical activity in the brain. As information passes into long-term memory, on the other hand, it is stored in a more durable form.

Neither the words nor the sentences are unduly long or difficult and there are no grammatical mistakes. So, individually, the sentences are easy to read but, collectively, they make us work too hard to follow their overall sense. In other words the paragraph is not fluent. Each sentence assaults us with new material without regard for what the sentences around have been telling us. We do not have the opportunity to get comfortable with the old material before we are asked to take in new material. Our minds have difficulty in pigeonholing the information in a logical way and this leads to at least two unfortunate consequences. The first is that we are obliged to store a lot of information while we back-track and re-read to find more clues about what to do with it. The second, as a direct result of the confusion, is that the material is likely to be interpreted by different readers in different ways.

Let us look at the bits of information that begin each sentence of this paragraph.

Memory can be divided into two phases: *short-term memory and long-term memory.* ***When an animal learns something*** *this information first of all enters the short-term memory where it will remain for a matter of minutes to hours.* ***The experimental methods used and the species of animal studied*** *can affect the precise duration of short-term memory.* ***A number of agents*** *including electro-convulsive shock (strong electric shocks applied to the head), low temperature, coma and deep anaesthesia can disrupt information that is being stored in short-term memory.* ***Any of these treatments*** *may produce a state known as retrograde amnesia, in which the memory of recent events is disrupted leaving earlier events unaffected.* ***Since more remote memories*** *are* *resistant to disruption, it has been concluded that the mechanism by which the information is stored in short-term memory differs from that for long-term memory. **Because short-term memory is disrupted relatively easily** by procedures which may be expected to have a profound effect on the electrical activity of the brain, it has been suggested that information is stored in short-term memory as reverberating electrical activity in the brain. **As information passes into long-term memory,** on the other hand, it is stored in a more durable form.*

In almost every case, the information that opens the sentence is either new or not related to the sentence before it. And here is the problem because the opening of the sentence is the very point where the reader is least prepared to receive and absorb new information. The sentences need to be made more 'user friendly' to allow the mind of the reader to tidy up and put away the material from the previous sentence and prepare for the next. Here is my attempt to apply this concept.

*Memory can be divided into two phases: short-term memory and long-term memory. **The short-term memory** is where information that an animal learns enters first, and this information remains there for a matter of minutes to hours depending on the species of animal studied and how it is measured. **Information stored** in the short-term memory may be disrupted by a number of agents including electro-convulsive shock (strong electric shocks applied to the head), low temperature, coma and deep anaesthesia. **Any of these agents** may be expected to have a profound effect on the electrical activity of the brain and disrupt the memory of recent events to produce a state known as retrograde amnesia. **However, retrograde amnesia** leaves the memory of earlier events unaffected. Since **memory of earlier events** resists disruption, the mechanism by which the information is stored in short-term memory probably differs from that for long-term memory. **So,** it has been suggested that information is stored in short-term memory as reverberating electrical activity in the brain and can be disrupted relatively easily. **On the other hand,** information that passes into long-term memory appears to be stored in a more durable form.*

This is easier to read because it now has a structure that presents new information only when the reader has been made ready to accept it.

You will notice that the first sentence has not been changed because, in the absence of any preceding material we have no 'old' information on which to build. In any case, it is an excellent topic sentence for the new paragraph, telling us what it is about.

But, in the second sentence, the new information giving details about the short-term memory is not raised until we have linked it with the old information from the first sentence. The modified sentence now begins by establishing that it is going to continue to tell us about short-term memory. Similarly, the new third sentence orientates us towards the now familiar theme of stored information before introducing new material about what can happen to this information. Similarly, the fourth sentence begins with 'agents' with which we are also familiar. The logical flow from one sentence to the next has been built up and continues throughout the paragraph. This has been done mostly by beginning sentences with words repeated from previous sentences but in the case of *However, So* and *On the other hand*, it has been done by using 'sign-post' words which also indicate clearly the direction of the sentence.

No new information has been added to the text, it is just easier for 'first–time' readers to grasp. By contrast, 'first–time' readers might have to read the original passage two or three times before being sure they have fully understood the whole message unless they are as familiar with the subject as the author. For example, now that you know what it contains, you could probably go back and read and understand the original paragraph without problem. However, most readers of scientific articles read because they want to learn and are not as familiar with the field as the author. Furthermore, even if they are familiar with the general field, they will read the revised text more rapidly and more easily, because they will grasp the details more quickly anyway.

Where to from here?

… you are probably too familiar with the work and what you wish to say about it to be able to judge its logic and fluency as objectively as it needs.

By now, the article should be approaching completion. It has a structure that is well thought through and is written in a style that that is fluent and easy to follow.

Or so you think!

You now have a problem that must be resolved. Your problem is that you are probably too familiar with the work and what you wish to say about it to be able to judge its logic and fluency as objectively as it needs. So, you need help.

The first place to seek it is among your co-authors. If you have taken major responsibility for the drafting of the article, then it is clearly their turn and their responsibility to contribute to the writing by helping you with the final editing. And, because you probably know more about the article than they do and what it is attempting to achieve, you should be directing them towards making a genuine and helpful contribution, not merely passing an opinion and leaving you to act upon it. There is nothing more frustrating than having co-authors returning a draft with no corrections of style. This clearly indicates that they have read it superficially at best or, at worst, not at all. To focus their interest, you could present them with a check list for editing like that on page 70 which, if they follow it, would compel them to think about and comment on all of the important issues in each section of the article. As a bonus, it can give both you and them a focus for a constructive discussion of passages about which you may differ for some reason. Without such a focus, you are likely to waste a lot of time. You will note that the suggested check list on page 70 covers many aspects of structure and style but says nothing about the quality and execution of the science behind the article—these are too diffuse for a book like this to cover specifically. In any case, they are the responsibility of researchers and should be addressed before and independently of the writing process. Nonetheless, they are paramount and will inevitably be a major part of any discussion between authors. So, it is a good idea to agree on issues arising from the science, but do so before discussing any differences in structure and style. To try to perfect the science, the structure and the style simultaneously is a big task and often results in confusion. So, in exercises of combined editing, it is preferable that you agree on each facet in turn; the science, then the structure, then the style.

However, even your co-authors may be so caught up in the familiarity of what they have been describing that they, too, may be assuming wrongly that what is clear to them will be clear to readers from all parts of the world. Ideally, you should make sure that at least one person, who has not been involved in either the work or the writing, checks that it makes sense and reads fluently. This is the 'colleague test', and is probably as close as you can get to assuring yourself that your article will be comprehensible to reviewers before publication and to the rest of the world afterwards.

The difficulty is that colleagues are usually busy with other things and don't have the time to devote to the review of a complete article in which they have no major interest except, perhaps, to help you. You can almost hear them groan as you throw a manuscript of, say, thirty pages on their desk and ask them to read it critically. They mentally estimate that they have a five-hour job, at least, ahead of them and, with the best will in the world they leave it there, intending to help you as soon as they have five free hours. Those five hours never eventuate, of course, and either the manuscript lies in their in tray for months or they return it to you after a cursory reading saying that it seems okay to them. Neither result is very useful to you. So, you need to be astute as well as considerate by asking them to do the job in small but well-defined segments rather than demanding a mammoth editing effort on the whole article. You can often do this as you go along. For example, if you asked your colleague to look at your *Introduction*, or even the outline of your *Introduction* with two questions: 'Does my hypothesis make it clear to you what I was looking for in this work?' and, ' Is the prelude to the hypothesis logical and make it a sensible thing for me to test?' Now, defined and constrained in this way, the task will take your colleague only a few minutes but will give you most of the information that you were seeking about this section anyway. At worst, it should stimulate a constructive exchange of ideas that will ultimately give you that information. You will have your response quickly and your colleague will be ready and willing to help again in a few days with similar, small but focused assignments relating to the *Results*, the *Discussion*, the *Summary*, the *Title* or any other part of the article.

... be astute as well as considerate by asking them to do the job in small but well-defined segments rather than demanding a mammoth editing effort on the whole article.

Final editing for style

To apply the final polish, you should run a last inspection and to be doubly sure you may even ask a colleague to do the same thing. In either case, you will be much more confident that things have not been missed if this final editing is systematic and not cursory. Sometimes, we can see a glaring lapse or a simple means of improving the text and, in our enthusiasm to fix them, overlook other faults. To help you pick up all of the avoidable blemishes, here is a simple check list of just five steps based on the information in this chapter. You will ensure that the editing is both comprehensive and thorough if you or your colleague follow it methodically. Edit the text, a paragraph at a time and ensure that each paragraph has been fully scrutinised before moving to the next.

Editing for style and fluency

Step 1. Is it a paragraph?

Check the first sentence to see that it defines the topic and the last to see that it is a genuine conclusion. Check the remaining sentences to see that they are relevant to that topic and are part of its development towards the conclusion.

Step 2. Do the sentences flow?

Examine the first words of each sentence and ensure that they include words that were used in the previous sentence or are 'signpost' words that relate the rest of the sentence to what preceded it.

Step 3. Are there stumbling blocks?

Check for words and expressions that may possibly distract the reader from the task of absorbing the message by causing doubt, introducing ambiguity, or needing several moments of contemplation to decipher.

Step 4. Can it be shortened without losing the meaning?

Remove expressions that don't add meaning like, 'our studies show …' or 'analysis of the data revealed …' Check if there are nouns that you can replace by verbs with the same stem and rephrase the sentences accordingly.

Step 5. Does it say what you want it to say?

The previous four steps are largely mechanical and can be done without considering closely the precise message you want to give in the paragraph. You should now re-read the amended paragraph to verify that it still says exactly what you wish it to say.

Choosing the journal

After considering and, where appropriate, acting upon the comments that you have elicited, you will begin to realise that your article is getting close to submission to a journal. This is the time to decide what is the most suitable journal. Some people advise that you choose a journal before you begin writing but this doesn't seem very sensible for two reasons. First, you will only have a general idea of what you want to say before you have completed the reasoning that allows you to write the article successfully. Certainly, you are unlikely to have enough detail to know precisely what you have to 'sell' to the world. Second, modern technology for word processing allows you to incorporate into your manuscript with little extra effort the 'house style' and other specific aspects that journals require.

Only after you know exactly what you have to offer can you decide with certainty the journal that is likely to have the most appropriate readership. The alternative, attempting to skew your article while you are writing it to fit what you suppose is a journal's readership, may lead you to play down the strengths of the article while trying to bolstering its less convincing aspects and this means ultimately that the article will be weaker than it should be.

Some authors and particularly some administrators have as their primary aim to publish in journals with very high 'impact factors', often abbreviated to IF. Indeed the administrators of many scientific institutions and grant-awarding bodies link their funding for researchers to the number of articles they produce and the impact factors of the journals in which they produce them. This is regrettable because, despite both of these measurements being quantitative and therefore easy to use for this purpose, they don't measure, or measure very poorly, the quality of the science and the purpose of the research—the two things that really matter. Much has been said and written about the value or the irrelevance of impact factors but they can have a devastating effect on the morale of young authors who seek, or are instructed to seek, publication in journals with the so-called highest impact. Some even talk of a strategy in which you start by submitting to the journal with the highest impact factor in the field and, if rejected, submitting to the next highest and so on until finally a journal somewhere down the hierarchy accepts the paper. Having a paper rejected, regardless of the reason, is demoralising at the best of times but to expose oneself or, worse still, one's students, deliberately to a high chance of being rejected several times is extremely injudicious. And the chance can be very high; some journals with high impact factors publish less than 10% of the articles submitted to them. Even protagonists of the use of impact factors point out regularly and rightly that the impact factor refers only to the journal as a whole and not to individual articles within it. So, to manage to publish in a 'high impact' journal an article that is likely to be read by very few of the readers of that journal may enhance the size of next year's grant but may do little to enhance the standing or influence of the authors among their peers.

Remember that the primary purpose of writing an article is to have as many people as possible read it, understand it and be influenced by it. This principle ought also to be the key to choosing the most appropriate journal for your article. Look for the journal that is likely to be read by the most people whom you would like to influence. You have already ensured that as many of these people as possible will read it and understand it by the way you structured the article and the fluent style in which you wrote it. So, you will have covered all of the essential objectives if you select journals that are read by people who read *your* type of article.

Sending to the journal

Many factors make up a 'good' scientific article and we have dealt with most of the definable and objective ones in this book. But there are subjective factors, too, that make an article 'good' for one reader and 'excellent' for another. One of the most important of these factors is the readers' backgrounds in research and therefore their possible differences in emphasis between the issues that are raised or their preference for certain words over other, equally relevant, words. That is why the perfect paper has never been written. So, don't keep your article circulating among colleagues for months or even years waiting fruitlessly for perfection. By the time you are receiving comments that are nothing more than minor differences of opinion and not of fact, or your own editorial changes are making no effective difference to the structure, style or readability of the article, you should send it to the journal you have chosen. After a final check of the references, verification that there are no typographical errors and that you comply with the 'house rules' of the journal laid out in its *Guide to Authors,* send it off.

Look for the journal that is likely to be read by the most people whom you would like to influence.

In doing so, you can increase your chances of a favourable reception by the editor of the journal by accompanying it with a well-planned covering letter that says more than, 'Here is a paper I would like you to consider for your journal.' When you chose the journal, you did so for a reason. For example, it may complement other recent papers in that journal or it may offer new insights into known problems that are in the journal's field. A statement, two or three sentences long, suggesting to the editor your motivation for thinking this journal may be the ideal medium for your paper could help the editor to gain a favourable first impression and give you a flying start. For the same reason, be sure to proofread the covering letter and check little details like the name and spelling of the journal and the editor!

Coping with editors, referees and reviewers

Now begins a justifiably anxious phase. For the first time you have no control over the process because it passes to unseen, unknown and possibly unsympathetic examiners into whose hands you have committed it. What happens while you are waiting? And are your anxieties as valid as they seem?

The first person to pass a judgement is, of course, the editor who will usually examine the title, the summary and the general layout of the manuscript to see if it is an article in the field that the journal usually covers and that it conforms with the 'house style'. This is where your well-crafted and targeted covering letter may be beneficial. You will usually be notified of the editor's initial decision within a few days. If the article does not appear to meet the requirements and scope of the journal you will be informed of this and advised to try elsewhere. If it is reckoned to meet the basic requirements, you will also be informed and told that the refereeing and editing process has begun. The editor will choose from an extensive list of potential referees, also called reviewers, two, or sometimes, three of whom seem to be actively working in your field. These will be sent a copy of your manuscript and asked to comment particularly on the originality of the work, the science, the methodology and your reasoning both for doing the work and drawing conclusions from it. They will usually *not* be asked to comment on anything but extreme breaches of style like spelling errors or incomplete sentences. That is usually the editor's job because the editor is more experienced in this field and because editors are the sole people who can control the uniformity of style throughout the journal.

Increasingly, journals are using electronic methods for speeding the transfer of information between authors, referees and editors but the refereeing process still takes time and patience, neither of which may be instantly available. Referees, being active scientists, are seldom immediately able to spend the several hours to review manuscripts properly and both editors and authors often become frustrated while manuscripts lie on desks awaiting attention for, possibly, several months. Irritated editors send them 'hurry-up' notices, anxious authors call up the editors to enquire about progress and the referees, themselves, feel guilty and overworked. In short, this process, called peer-review, is far from ideal. But it is by far the best way we have to ensure that what is eventually printed is good and acceptable research that is sanctioned by 'the general research community'. Anything less than peer-review risks putting the whole of research into disrepute by attributing dubious science the same status as good science. Almost every experienced author has suffered, at some time or another, a long and exasperating hiatus between submitting a manuscript and having it accepted for publication because of a sluggish referee. So, if you chance on such a referee on your first attempt at publication, be philosophical and accept that all of this is part of an unfortunate but very necessary step in the research process.

The editor's second and most critical appraisal of your work comes after they receive the referees' reports. Based on these reports and a re-reading of the manuscript, the editor judges the acceptability of your work for publication.

If all referees say that the work is good and the editor thinks that the style and layout are fine, you will be sent a letter, on paper or electronically, saying that the manuscript is accepted and, with a few minor corrections, will be published in a future edition of the journal. If you get such a message, you have reason to celebrate because that sort of letter is, perhaps surprisingly, rare these days. More common, but still good news, is notification that the referees or the editor, or both, propose that the paper should be modified for various reasons and that, if you care to address their recommendations, the editor will reconsider the paper.

These modifications are usually classified as 'minor' or 'major' based on the amount of work that the editor estimates you will need to do.

You are bound to be initially disappointed or even angry that anyone has found fault, whether minor or major, with the work that you reasoned so well and crafted so carefully. That is why it is often a good idea to put the editor's response in a drawer for a day or two while you calm down lest you send off a hasty and ill-considered reply. In coming to terms with their comments, it is well to consider the roles that editors and referees play in the publishing process. Neither of them is there to ensure that your paper is rejected as you may be tempted to think. Editors have two main tasks; to produce regular issues of their journal and to maintain its scientific and literary quality. To do so, they have to have articles from authors like you but they want your work to be of the highest standard that both of you can achieve. They usually assume total responsibility for the literary and formatting standards but, because they are not usually experts in the detailed field, they choose referees to help them assess the scientific merit and integrity of the paper. The referees they choose are usually active researchers in a field close to that of your article and they are asked to comment on your methodology, results and scientific reasoning. The editor does not automatically consider the referees as being any more competent than you, but simply as peers who will judge the work from a different point of view. So, if you disagree with a referee you should be able to convince the editor with a logical argument why the segment of your article that has been criticised should remain and you will not have to make the suggested changes. Similarly, if a referee says that your paper should be rejected for some reason but the editor indicates that he or she is prepared to look at a revised version, you live to fight another day. Only the editor has the power of rejection and the editor's covering letter is the crucial document in the material that comes back to you after the refereeing process.

If that letter says that your article has been rejected, then you should not waste your time or your reputation trying to get it reconsidered by that journal. Use the experience and the comments to make a realistic judgement about whether to improve your article and try another journal that may consider it to be more appropriate. Remember, the peer review process is better than any other checks-and-balances process that we have, but it is still based on human judgement. That means it has a strong element of the well known human failings of subjectivity, opinion and, dare we admit, error and abuse that could result in one or more of the two or three referees and, finally, the editor getting it wrong. You can easily bolster yourself in your disappointment by remembering that there have been many chronicled examples of pivotal scientific findings that have been rejected by several reputable journals before being finally published and making their impact on the world.

Re-submitting to the journal

If you have been invited by the editor to re-submit your article for consideration, do it carefully and you have a very good chance of success. In fact, by inviting you to re-submit an amended paper, the editor has told you as much, and the good news is that you now have a written set of conditions to follow. But, you must follow them diligently and completely by addressing every point, however small, made by each referee and the editor. For each point, make a note of your action and, where you have not followed their advice, partially or fully, record your reasons. Then, make a separate list of all these responses and include it with the modified manuscript when re-submitting your article to the editor. This makes it easier for the editor to check your changes and to judge whether you had a good reason not to make some of the recommended changes.

While addressing the changes that have been suggested to you, you will smooth the way for the second passage of the article if you mix a large slice of diplomacy with your natural desire to argue that you are right and the referee is not. In any case, and unfortunately for your ego, you will find the referees generally have a point and you can accept the criticism and make the alterations knowing that it has improved the manuscript. But, from time to time, you will find comments that fit the category of a whim or an opinion and, in your own judgement make little or no improvement to the work at all. This is where you must ask yourself whether you could live with the referee's amendment even though you do not find it any better than your original text. If you can, be pragmatic, make the change and thank the referee for the suggestion. By doing so, you build up your credibility as a cooperative author and not as one who carps at suggestions most of which were probably made in good faith. This credibility is important, because your protest will be taken more seriously by the editor when you cannot agree with a recommendation that you believe will compromise the scientific story that you are telling. In short, it is to your advantage to get rid of any suggestion that your case may be based on pique or anger by accepting marginal and unimportant suggestions with good grace.

… the good news is that you now have a written set of conditions to follow.

Good editors, in considering an author's reasoned argument for rejecting referees' advice, do not deem referees to be cleverer than authors simply because they are referees. Instead, they weigh up the cases for and against change and make a judgement on those alone. So, if you were told,

Your Introduction lacks substance because it does not mention the important work of Bloggs (2007),

and you replied along the lines:

Despite its importance in other ways, Bloggs' work is not concerned with the justification of the hypothesis I tested in this paper. It would be distracting to the logic if I included it,

most editors would accept your argument without further question.

Finally, the future of your article depends on your response to the editor, so make it good. Be meticulous in addressing even the smallest of points raised by the referees or the editor and, equally meticulously, prepare a document listing how you have handled each point as suggested above, so that the editor can use it as a check list. Return the list with the revised manuscript and a brief covering letter as soon as you can, but give yourself time to reflect on your responses, especially to any controversial issues. Remember that, by inviting you to re-submit and defining the conditions that would make the article acceptable, the editor has made a tacit commitment to accepting it. Your job is to make it as difficult as possible for the editor to find a reason for not sticking to that commitment by addressing every condition that has been raised.

Thinking and writing beyond the scientific article

RESEARCH THAT YOU DO INTERESTS A WIDER RANGE OF PEOPLE than just those who read the research journals. Communicating research orally, in posters, as reviews, as articles to inform and interest non-scientists or in theses widens your sphere of influence. But for each of these media there is a different objective that calls for an approach often radically different from that you would use in writing for a journal. In this section, we look at these objectives and the ways of coping with them to communicate successfully.

The text for oral presentation at a scientific seminar 78

Design and preparation of posters for conferences.................. 88

The review 95

Writing science for non-scientists 100

The thesis 106

The text for oral presentation at a scientific seminar

If writing your first scientific paper appears to you to be a daunting task, then presenting a paper at a scientific conference is likely to be even more scary. Not only are you presenting your data to be scrutinised by the audience, but you are doing it in 'real time' and you are presenting yourself as the living embodiment of the work you are putting forward. So, even if your data and reasoning are sound, they run the risk of being overlooked if you make a hash of telling people about them. One approach that many speakers adopt to minimise their risk of making gaffes is to try to emulate other speakers who seem to have survived unscathed. In this way, they can't be singled out as being unusually naïve or irritating. This is not a bad idea but it depends heavily on whom they are attempting to emulate. If they decide to copy people who are neither naïve nor irritating then they won't necessarily be unusual but they may still frustrate or have no impact on the audience. In fact, a far better approach is to seek to be unusual, but to be unusually good. A presentation that avoids the conventional clichés and the predictable, but uninformative rituals and sets out to produce a fresh, and enthusiastic performance will encourage people to take in your new information and to retain it.

... seek to be unusual, but to be unusually good.

A performance? Most certainly! What else describes a solo presentation lasting from 10 to 30 minutes on a stage in front of a group of, maybe, several hundred learned people? Of course, there is the need to have a script that is meaningful and worth the audience's attention. But without skilful and perceptive delivery, even good data can be overlooked by participants at a congress who are in the process of being overloaded with information. Some people find performing comes more naturally to them than others but, with some care and some guiding principles, everyone can achieve a standard that is both creditable and effective.

These principles can be broken down into those that apply to the *structure* of the presentation and those that apply to the *style*. Of the two, the s tructure of the presentation is still the most important and, in many cases, the least understood. A well-structured presentation can compensate to a great extent for a lack of natural flair or flamboyance in the style of the presenter.

Structure

We have already seen that a well thought out structure is a key component in written scientific papers. In papers for oral presentation, it is still essential but very different. An oral presentation needs to be structured to meet four principal objectives.

- To get the attention of the audience and hold it.

- To get the audience to remember at least something of your message.

- To get them to remember the right bits of your message and not the wrong ones.

- To complete the talk before you are gonged off!

Getting the attention of the audience

Your message is the *only* thing that you want your audience to concentrate on: don't distract them with anything else. Flashy technology is to help you get your message across—not to impress people in its own right. But ... sloppy presentation also draws attention away from your message.

The classical structure of a good talk is an opening that says what you are going to say, a developmental phase that gives the details of what you want to say and a finale that summarises what you said. This time-honoured sequence works because the first part takes into consideration the need of your listeners to have a broad expectation that acts as a context into which they can absorb the details you want them to retain. The second part gives them those details and the third part gives them time to reflect on the whole presentation by reinforcing your key message.

When preparing your paper it is a good idea occasionally to take a pessimistic view of how the audience might be thinking at the time you will be presenting it. Imagine, for example, that you are the third speaker in the session after lunch and you have been preceded by two particularly boring presenters. The audience, despite its best intentions, is bored, sleepy and looking for distractions. You have been introduced by the chairman and arrive at the speaker's desk. It would be unrealistic to think that, in such a situation, the audience would now be waiting eagerly to gather every pearl you are about to cast before them. Instead, they may be thinking that your hair is untidy, you look nervous, the room is too hot, or the session is too long. Here is your first big hurdle. You could make a spectacular opening by tripping over the microphone cord or knocking the water jug into the chairman's lap, but this would be a hard act to sustain for the rest of the talk. It would certainly unify their thoughts but not in a way that would encourage everyone to take notice of what you are saying. Words are your main equipment and your opening sentence is crucial. It must make an impact and, at the same time, make people wish to hear more. Don't waste it with an opening such as:

The title of my talk this afternoon is, The effect of Leukemia Inhibitory Factor on synthesis of milk protein in bovine mammary epithelial cells. A title is not a statement of what you are going to say but a statement of the subject that you will be addressing. These are not the same thing. If the person chairing the meeting is doing a proper job, the audience will already know the title of your talk anyway.

Or, you could immediately launch into a detail such as:

Smurch et al. *(2006) in extensive studies into bovine mammary epithelial cells concluded that their production of milk proteins may be inhibited under certain conditions ...*

You cannot expect listeners to understand the 'little picture' if you haven't first given them a chance to see the 'big picture'. All but the most ardent listener will be put off by a beginning like this that launches directly into details. How much more animating (at least to an enthusiast of myoepithelial cells) would be :

This afternoon I am going to show you that Leukemia Inhibitory Factor inhibits milk and protein production in the mammary gland of the cow and by suppressing its activity we can enhance the production of milk.

Apart from giving more information, this opening sentence tells the listener what to expect. It is a résumé of the whole presentation in one sentence.

You cannot expect listeners to understand the 'little picture' if you haven't first given them a chance to see the 'big picture'.

So, before you start to write your talk, form yourself a vision of the 'big picture'. That will be your main message. Now, think of the most spectacular, or impressive, or thought-provoking thing that you can *honestly* say about it.

This is your opening statement.

But, have you revealed too much of your talk by adopting this approach? Not at all. Your function is to get across a message, not to have secrets, spring surprises or recount a complex mystery. It may seem blunt and unsubtle, but the adage 'Tell them what you are going to say, then say it, and then tell them what you have said' works wonders in scientific presentations. For our purpose the adage can be translated: Make an arresting beginning containing a 'micro-summary', then use the body of your paper to present evidence and convince the audience, and then round off with a conclusion containing the 'take-home message'. Not surprisingly, the 'take-home message' is exactly the same as the message with which you opened. So, if you think that you have problems in deciding how to finish off a talk, your problem is solved the moment you decide on how to open it. The key moments when the audience will be most attentive are the beginning and the end of your talk. Make them both count by carefully crafting your most compelling message for these moments.

Get the audience to remember at least something of your message

Having gained the attention of all of the people likely to be interested, how do you keep it? You are talking in 'real time' to a living, breathing and thinking group of people. They will continue to live and breathe regardless of what you say but, if they aren't thinking about what you have to say, you might as well be speaking to a brick wall. So, you have to get them to feel involved with the work you are presenting. This requires that you prepare your paper in a style that converses with them rather than just conveying information. A conversational style, contrary to some views, does not reduce the scientific merit of the paper. The magic word that helps produce a conversational style and which you ought to

SCIENTIFIC WRITING = THINKING IN WORDS

use as often as seems sensible, is the word *'you'*. Whenever you use it listeners feel that they are participating in the paper.

You may wonder why we used …

If you look at the two numbers on the right hand side of the table you will notice …

The slope of the line is not as steep as you might expect …

Each time you use *'you'*, your listeners, whether they like it or not, are encouraged to make certain that they are not being maligned, misinterpreted or otherwise taken in vain and, in so doing, automatically pay attention. To a lesser extent the word *'I'* (or *'we'*, if appropriate) and the active rather than the passive voice achieve a similar result. They involve the speaker with the substance of the talk.

I (or we) couldn't get two sets of data because …

instead of *There were two missing sets of data …*

I interpret this to mean …

instead of, *This seems to mean …*

This personal style helps your listeners to picture you, the person they see in front of them, working your way through your experiment or your reasoning and this reinforces the message you are conveying.

Another way of retaining attention is to include some humour in your talk. It lightens a heavy session and if you can successfully incorporate a joke in the early part of your talk, you will keep at least some people's attention if only because they are waiting for another. Unfortunately, not everyone tells jokes effectively, and not all audiences respond predictably to humour. Laboured humour is worse than none at all. Anecdotes that commence with 'Did you hear the one about …' or, 'That reminds me of the butcher with one leg …' should be avoided because they announce that you are about to tell a joke and imply that you expect them to laugh at the end of it. This can sometimes set you up for a very uncomfortable downfall. Instead, successful speakers at scientific meetings develop their punch lines as very slight variations of the serious text and rely on unexpected turns of phrase rather than pre-announced jokes. In this way, the humour wastes little time and gains a great deal of attention—if it works. If it doesn't work, as occasionally happens with dull or sleepy audiences, you can continue on without even revealing that you expected a response. If you have set yourself up with an introduction like 'Did you hear the one about …' you can't avoid being acutely embarrassed when the audience doesn't respond to the punch line.

Recognise from the beginning that you are smarter than your audience

This may seem arrogant but it makes sense for two reasons. First, why are they here to listen to you if you are not smarter than they are? Second, even if you think that there are people who are cleverer than you, you are still the most competent person to talk about your data. You are the one who spent months or maybe years working on your subject in your small corner of research and most of the audience will be getting much of this new knowledge from you during the few minutes of your talk. Hence, it is imperative that you give them every chance to keep pace with the flow of information—but how much time do they need? Experts tell us that a dedicated listener who has no prior intimate knowledge of a field is capable of absorbing one new idea every three minutes. If ideas are presented to them more rapidly than this, they lose concentration and stop absorbing new information.

When structuring a presentation scheduled for 30 minutes, this translates into confining yourself to 10 new thoughts in the entire presentation. On the other hand, if you are to talk on the same subject for only 10 minutes you must restrict yourself to just three separate ideas and abandon the idea of introducing the other seven. This is often hard for an enthusiastic scientist to accept. There is so much to be said; wouldn't it be better to raise each of the 10 ideas very briefly and so put on display the full scope of the work rather than leave out whole aspects of the subject you want to cover? The answer is emphatically, No. Such an approach inevitably creates a gridlock in the listeners' brains and they absorb and remember nothing. By contrast, three major points, fully and convincingly presented, will have a much stronger chance of leaving a lasting impression and, with some listeners at least, motivate them to look further into the subject. And that may well include most of the other seven issues that you decided deliberately to set aside. In fact, some of these points are likely to come up in the time left for questions if you have successfully interested your audience.

One successful way of summarising and, at the same time introducing variety into your style is to alternate regularly between the general and the particular.

Because you are so familiar with your field you may think yourself boring when you slow down your presentation to give your listeners the time to catch up. You certainly would be boring if you simply repeated exactly the same message over and over for three minutes. But if you cleverly approach the same subject from different angles, you will allow the full impact of what you are explaining to become progressively apparent to the audience. Some angles will be more successful with some listeners than with others because they will be associating your information with different information that they already know. After all, association with what we know is the way that most of us learn and retain information.

Fortunately, in a spoken presentation, you have much more liberty to repeat yourself than in a written one. In fact, because readers of written articles can check things out as they go and pace the rate at which they absorb your information and listeners cannot, you are obliged in a good oral presentation to help the listener to catch up. So, you can summarise as you go. Your summary will allow the listener to reflect on what you have just said without feeling that they may be missing what you are about to say. One successful way of summarising and, at the same time introducing variety into your style, is to alternate regularly between the general and the particular. A series of details can become dull unless they are broken up by a generalisation that sums up what all the details mean. Similarly, broad generalisations can be enhanced and clarified by giving specific examples, especially ones to which a listener is likely to relate, to illustrate the impact of those generalisations As you talk you will most likely have some visual material to help you, electronic slides, pictures, graphs and tables or sometimes interesting physical material that can act as a dramatic visual demonstration. Each of these allows you to reinforce the point that you are making and to give the listener time to draw level with your thinking without your being obviously repetitive.

Experienced presenters become skilled at 'reading' the response of their listeners by observing closely their responses to what is being said. When they notice that attention is beginning to waver they use this as a signal to move to the next major point. When they reckon that listeners are still coming to terms with the current information and need more time, they slow down to allow that time before moving on. Slowing or speeding your rate of delivery—within reasonable limits, of course—can keep the pace of your presentation in unison with the capacity of listeners to follow it.

No one will criticise you if you read—only if you seem to be reading!

All of this may suggest that your presentation should be an *ad lib* affair, delivered according to your assessment of the audience rather than according to a pre-planned structure. Not at all. If your talk is to have a good chance of being successful, it must be carefully structured and presented as closely as possible to that structure. Minor adjustments to suit the mood and the aptitude of the audience are fine, but whimsical flights of fancy that come to mind for the first time in mid-presentation are potential recipes for catastrophes. You can never be sure of how long they will take to get across and you will have even less idea of the effect they will have on the listeners. And once you leave the structure you have carefully set for yourself, it can be a nightmare trying to recover it again. Even if you have admired a particularly gifted presenter who appeared to have no notes at all and who seemed to improvise the whole talk, do not be deceived. If the talk was a good one, it had probably been prepared and rehearsed over and over to polish it to the level that you finally witnessed.

So, stick to your script, but remember that the impact you make will be lessened if the audience notices you obviously reading it and thinks that you depend on a prepared script. One strategy is to rehearse the whole talk and commit it entirely to memory so that you go to the rostrum to deliver it without any obvious notes at all. But, this can be dangerous because, without back-up, you can be very vulnerable to even minor lapses of memory or concentration. A safer option is to take your notes with you. No one will criticise you if you read—only if you seem to be reading! The key therefore is to have your notes to help you but make sure that you do not appear to be reading them. That is not as difficult as it may first seem if you follow three simple guidelines.

1. Rehearse your talk so that you know what is in your notes, if not by heart, then well enough to pick up prompts from a few words rather than needing to read the complete notes.

2. Never read and speak simultaneously. Try to spend every moment that you are speaking making eye contact with the audience. When you read, read in silence.

3. Time your periods of reading to coincide with either the 'dead time' after introducing a new overhead or the end of the delivery of a piece of information that the audience will need a few seconds to digest. Your audience will interpret your silence as good manners or skilful timing. The chances are that they will be so preoccupied with their own thoughts that they will not even realise that you are reading the main elements of the next part of your delivery.

Another option is use audio-visual aids, not only to help the audience but to act as prompts to keep you on track as well. In fact, some audio-visual software allows you to have prompts on the screen that you see but not on that seen by the audience. However, the same guidelines apply. Do not be caught reading text verbatim from the screen. It is even easier for the audience to catch you out than when you read from notes on a piece of paper. By all means, make what you say reinforce what you have placed in front of your audience to read, and *vice versa* but use different words and, preferably, a different approach. In any case, the material on screen should be brief dot points or headings, certainly not long tracts of text, so you will automatically be obliged to elaborate on them in your own words and so appear spontaneous.

Dead time

We normally think of an audience as a group of people who are there to listen to what we have to say. However, the moment that you present them with material on a screen you are asking them to do two things at once, listen and read. In fact, most people cannot satisfactorily do these two things simultaneously and you should keep this in mind when constructing your presentation. So, if you want people to take in what you say, the material in overheads or slides should be as brief as possible. Once a new piece of visual material arrives on the screen, it automatically becomes the newest and most absorbing thing to which the audience turns its attention. The amount of time before the audience returns to concentrating on your spoken word depends largely on the amount and complexity of this material. This time is often called 'dead time' because whatever you say within it is mostly or wholly ignored. Try the experiment sometime—before a friendly audience, of course. Put up an overhead and then deliberately look at the audience without speaking. You will find that it is five to 10 seconds before the first people begin to register that they are reading in silence. Alternatively, say something totally absurd during the 'dead time' and the chances are that no one will notice! You can use the dead time to read your notes in silence without being noticed or, if you, rather than the audience, are uncomfortable about being in silence, talk about incidentals on the slide which the audience will be looking at anyway such as the axes on a graph or headings in a table. Certainly, when structuring your presentation, remember that this is not the moment to deliver key messages in your talk or even the key messages that the slide is illustrating. Wait until the audience is back in listening mode.

A paper for reading and a paper for speaking

The content of an oral presentation of some research at a conference is very different from that of a written article for a scientific journal about the same research. So much so that one can never substitute satisfactorily for the other. This means that, if the conference at which you are to talk also publishes a *Proceedings*, then you will be obliged to write two articles , one for the *Proceedings* and the other for your presentation. They will differ in almost every aspect except the subject matter, as the table below summarises.

How a paper for an oral presentation at a conference differs from a paper written for the proceedings

Component	Paper for presenting at a seminar or conference	Paper for the proceedings of the conference or seminar
Structure		
Opening sentence	Vital that this sentence has the most impact possible.	Can be less conclusive and more introductory.
Introduction	40% of total (time).	5–10% of total (space).
Methods and Results	40% of total (time).	40–60% of total (space).
Discussion	20% of total (time).	30–60% of total (space).
Closing sentence	A clear résumé of the most important message—similar or complementary to opening sentence.	Resounding closing sentence not necessary.
Summary	Provide mini-summaries throughout the article.	One comprehensive summary in a special section.
Style and layout		
Repetition	Highly desirable.	Very little.
Length	To finish just before time.	As short as possible.
Accessory material	Slides or Power Point or similar as reinforcement of text.	Only those tables and figures that are relevant.
Humour	Desirable but not essential.	Undesirable.
Grammar	Sound, but minor lapses of grammar forgivable.	Sound, with impeccable grammar.
Style	1st and 2nd person used often. Conversational and simple.	1st person used sometimes; 2nd person never used. Precise, clear and brief.
References	The least possible.	The required number to support sound arguments
Acknowledgments	The least possible.	Brief but adequate.

Timing your talk

All researchers consider it an achievement to be selected to give an oral presentation at a big meeting. They are justifiably excited to have the chance to expose their work to colleagues but they should not let their enthusiasm push them to try to say too much.

Nothing spoils a good talk so completely and predicably as going over the allotted time. To do so is not only discourteous to following speakers, who could be forced to alter their presentations to fit into less available time, but it gains the speaker no advantage whatsoever. Speakers who think that they have profited because they have crammed in a few more facts when they steal an extra minute or two, should think again.

In fact, one of the first rules of chairing a meeting is to run the session precisely to time and good chairs are ruthless to the point of being rude with speakers who try to gain such an advantage. They usually do not allow them to continue even a few seconds beyond their allotted time. The tragedy is that the part of the presentation that is cut off is the final, concluding, take-home message—the part that the audience is likely to remember. And this is, of course, the most vital part of the whole talk. As a result, the whole talk will have no readily discernable structure and inevitably will be a flop. But, what if a speaker finds an ineffectual chair and ploughs on to finish the presentation despite being asked to stop? The result is the same. From the moment that the chair begins to approach the speaker with the clear intention of closing the talk, the audience recognises that it has a new spectacle to enjoy. It becomes absorbed with how the chair will act and how the speaker will react. Will there be an argument? Will the speaker pretend to ignore the presence of the chair or the chair's instructions? How? The important point is that whatever new information the speaker manages to deliver during this time will be wasted on an audience that will be in no mood to listen to it. A disorganised scramble might amuse them but for entirely the wrong reasons.

So, the case for sticking to time is overwhelming and there are simple ways that you can go about it. First, calculate the maximum time you have available. This will be 10% less than the time allotted in the program, minus another 30 seconds. In other words, for a 10-minute talk you should plan to complete the presentation in eight and a half minutes. The 10% is to account for delays, the introduction by the chair, slight adjustments in time because of public announcements, 'dead time' following each presentation on the screen, minor glitches with the equipment, for example fitting lapel microphones or bringing up your file on the computer, and other contingencies. The 30 seconds is to cover the fact that most people generally speak more quickly when rehearsing behind the bathroom door than in a large auditorium in front of an audience. If, in doing this, you overestimate the time and finish early, nobody except the chair will notice because they will not be timing you but, at least, you will not offend the audience and you will always be appreciated by the chair. On the other hand, if you underestimate the time and risk going over time even by a few seconds you could ruin your whole presentation.

Having decided on exactly how long you need, rehearse and time what you have to say. If you find that you are taking longer than your estimated time, mercilessly wipe out what you consider to be the least important information you were going to present until you get inside your time limit. When your presentation relies on slides or a PowerPoint presentation, the safest way to gain time is to remove slides. As a rule of thumb, you are likely to spend between one and two minutes on each slide. If you spend less then this, you will be probably speaking too fast and, if you spend more then two minutes, your slide is probably too complicated. In any case, if you have more than six or seven slides for a 10-minute talk, it is a signal that the talk may be too long.

As a further precaution, write out your closing statement of one or two sentences—the take-home message—and learn it by heart. This will give you the opportunity, if you need it, to launch into this, the most important part of your talk at any time. It can be a lifeline if you have somehow lost your way in the body of the talk and spoken at greater length than you planned, leaving insufficient time to complete it. In your notes, highlight a paragraph near the end that you would be prepared to delete to make time for the all-important finale. Even though you may feel that deleting it might spoil the flow, that will be far less obvious to the listeners than if you were obliged to cut your presentation short without an ending.

Leaving out material in order to arrive at a forceful and dignified finish to your talk may sometimes be hard to accept but it is not as drastic and disappointing as it may first seem. For example, there is generally a question time at the end of most oral presentations. This can be used to fill in details that you were obliged to abandon, should some member of the audience ask. Remember, too, that conferences and seminars are seldom confined to formal presentations. There are generally many opportunities for people who have been interested by your talk to consult you privately on your presentation or other aspects of your work. The essential requirement is for your paper to be stimulating and interesting in the first place. That is why your first priority must be to present your main ideas effectively and leave a good impression.

Ending your talk

Recently, there seems to be a predilection for presenting a final slide or overhead offering acknowledgement to all and sundry who had anything to do with the work that was presented, or one that has a funny picture with words like 'Thank you' or 'Thank you for your attention'.

Think about this for a minute. You have striven during your presentation to get across a coherent story and have concluded with the most important conclusion from this story only to sweep it away and replace it with a list of people the audience is unlikely to know or care about, or a dubious statement of gratitude because they kept quiet while you talked to them. After all, it is only at political rallies that they would be likely to heckle you if you had done a poor job.

In many cases the last slide is left on the screen during the whole of the question time. What a tragic waste of an opportunity to reinforce your message as the listeners catch up in their minds with what you have just presented to them. Indeed, a well-presented message in front of them might stimulate some interesting questions from the audience. By contrast, I have never been aware of a list of collaborators stimulating anything.

The purpose of any well-conceived presentation should certainly not be to introduce a list of collaborators and certainly not make a feature of them. It may make you feel virtuous but, to your listeners, it is simply an unwanted piece of trivia. However, if, despite this, you feel that you must acknowledge somebody, do it in the body of the presentation as an aside and at a time when you feel that the audience needs a moment or two to catch their thoughts. For example, you could show genuine appreciation for the input of a colleague or group of colleagues when you are describing a particular piece of work or reasoning in which they were involved. But your final, general slide should be left as a vehicle to reinforce the message that you are delivering.

Design and preparation of posters for conferences

The poster acts as a catalyst to stimulate members of the audience to communicate with the author or authors for as long or as short a time as is comfortable for both of them.

In the last 25 years, a new form of scientific communication—the poster—has developed and become widespread. At large conferences, scheduling oral presentations into the proceedings for everyone who wanted to make one became increasingly difficult. Some conferences were running six or seven simultaneous sessions to allow all participants to speak. This was not only costly but inefficient because attendees often wished to hear two presentations scheduled for the same time in different venues.

So, a popular solution was to have at least a proportion of the presenters prepare posters summarising their work, pin them to boards in a special room or rooms for the purpose and be prepared to discuss the work with people that were attracted to them. Whole sessions of conferences are now set aside for the display and discussion of posters and delegates arriving with cylindrical rolls containing their contributions to the poster sessions are now a familiar sight at the registration desks of most modern conferences.

The traditional formats that are familiar in written articles or even in oral presentations are, of course, of little use in the preparation of posters. The medium is relatively new, and presenters have been experimenting with alternative formats so that the principles that characterise a successful poster are only now beginning to emerge.

There are two major differences between posters and oral presentations. The first and most obvious is that the audience for posters is not captive. People can stroll through tens or even hundreds of posters, choosing to read whichever they wish and ignoring completely those that they find unattractive. As a consequence, delegates at conferences can become overwhelmed with the amount of potential information in front of them and miss seeing or responding to many presentations within the whole mass available to them. The second difference is that oral presentations fill an allotted span of time whether they are interesting to the audience or not. Posters can hold the interest of individual members from as little as few seconds to an hour or more and the material covered can extend well beyond that covered in the title or the contents. The poster should act as a catalyst to stimulate a member of the audience to communicate with the author or authors for as long or as short a time as is comfortable for both of them. The oral presentation is strictly confined to the advertised subject matter and the advertised time. For these reasons both forms of presentation have their enthusiasts and detractors but it is certain that posters now have a permanent place in modern scientific communication.

What makes a successful poster?

A poster has four objectives. It must, in sequence:

- catch the eye
- make a statement that arouses the scientific interest of a passing onlooker
- provide justification in the form of data, and
- stimulate the onlooker to find out more by talking with the author.

Not only must it meet these objectives but, for at least the first three of them, it must do so within a very short time. Complying with this rigid time constraint is the key to designing successful posters.

Imagine delegates to a typical conference entering the poster hall to view and become informed by as many as 200 posters in no more than, say, two hours. If they perused all of the material in front of them they would have an average of just 36 seconds to devote to each poster. To do so would clearly be impossible so they have to choose which ones to ignore and which to explore thoroughly. They start to stroll down the rows, glancing at each poster to determine whether it will be worth a second (or possibly a third) glance. In well-organised conferences, delegates may be supplied with sufficient preliminary material for them to do their 'poster shopping' before the session, or even before the conference begins. Even so, the delegate may still plan to see a long list of posters and will always be prone to being distracted by an appealing interloper, so catching the eye is always important. *Each poster has about 2 seconds to catch the eye of each delegate.*

If it is successful at catching the eye, the delegate will begin to search for something of interest. There are three possibilities here.

- The poster is not in the field of interest of the delegate who determines this rapidly and moves on.
- The delegate cannot decipher a message from the poster rapidly and makes a decision not to waste more time seeking further and moves on.
- The delegate finds a key message that the poster is about something that is interesting and stays to read the rest of the poster in detail.

Each poster has about 10 seconds to stimulate scientific interest.

If the poster has succeeded at this hurdle, the delegate—now a reader—starts to seek justification in the poster for the statements that made it originally attractive. This justification and elaboration will come in the form of data or statements of detail that expand on the primary information. *Each poster will be read for 30 to 60 seconds.*

Now, the reader will be sufficiently informed about the work on show to be able to ask the author questions about methodology, details of other work, planned or already done, prospects for employment or anything else imaginable.

At this point the poster will have successfully accomplished its mission. *There is virtually no time limit on discussions between interested readers and authors of posters.*

The structure of a successful poster

One of the saddest sights at conferences is authors of posters standing forlornly and alone beside their work, because they have failed to take heed of the four objectives outlined above. In view of the incredibly short constraints on time for a poster to achieve its objectives, it is clear that good posters must follow a format that is radically different from that of a scientific article for publication or a paper for oral presentation. The time-honoured sequence of *Introduction, Materials and Methods, Results, Discussion* and *Conclusion* simply fails to work at the level of the poster. The objectives are different so the format must also be different.

... good posters must follow a format that is radically different from that of a scientific article for publication or a paper for oral presentation.

Catching the eye

We sometimes see posters that show the hand of professional graphic designers and these invariably attract attention in the way that well-designed advertisements do. Many authors of scientific posters do not have access to such specialist help and have to resort to less professional and usually less expensive means to be attractive. This should not be a major setback. After all, simply catching the eye may be important, but is not what induces other scientists to stop and read the work. The content and the skilful presentation of the content are what does that. In any case, most scientists have sufficient flair to ensure that the layout of their work is pleasing enough to ensure a second look from most passers-by. In an earlier chapter on written papers, when we looked at ways of emphasising the important points, we found that, apart from position and size, there was little scope for making certain things seem more important than others. Posters do not have most of those constraints. The choice of colours, or even the presence of colours to replace dull white, the distribution and content of photographs and figures, the use of attractive fonts of a variety of sizes and the imaginative use of diagrams can all combine to lure the viewer to take a second look.

But, before this, we must be aware of a number of things that definitely make posters unattractive and get rid of them. The single most common fault of posters is an oversupply of information. A presentation that looks like an oversized page from a textbook has no visual appeal and, because it is obvious that it cannot be read in 30 to 60 seconds, most people, anxious to look at all of the posters on offer, will not even attempt to do so. Another 'turn-off' is those unimaginative posters that consist of a number of sections, usually those that are found in written articles like *Introduction, Materials and Methods, Results* and *Discussion*, each printed on A4 paper and pasted onto a board in the allocated space. Even worse, is text or information that has no direct bearing on the message that the poster is supposed to be delivering. These make reading a daunting task when standing in the middle of a room among several hundred participants at a congress. A third widespread failing among designers of posters is to present them in a font that is too small to be read from further than about a metre or so. In none of these cases does the work have a reasonable opportunity to catch the eye of passers-by.

Making a statement that arouses scientific interest

This is the part where some hard decisions have to be made. You must condense all of the scientifically important information in your presentation to words that can be read in no more than 10 seconds. This means, effectively, about three sentences that will be prominent enough to be taken in during the reader's first scan of the poster. The sentences do not have to be all together and they may be supported by other, less prominent, material to which the reader may return later, but they must be the distilled wisdom of the presentation. In other words, they can be well placed and well constructed headings or a couple of sentences that sum up the whole story that you are telling. With this simple criterion in mind, it is clear that there are several components of a traditional, written paper that we must completely ignore. As we have seen, the headings, *Introduction, Materials and Methods, Results* and *Discussion*, are almost invariably inappropriate for a poster. The headings themselves convey nothing new and people simply do not have the time or the inclination to read through the material underneath them. Their function in a written paper is to serve as signposts for the reader to locate specific sections. In a poster, you can replace each of these headings with the sentence or two that best summarises the whole section. And the order in which you present them can be quite flexible. In effect, the passing reader is usually seeking the major results expressed in summary form coupled with the major conclusions. Unless the poster is about methodology, it is seldom sensible to bother at all about materials and methods at this level, certainly not in any detail.

… keep in mind that the purpose of the poster is not to tell the full story, but to induce people to talk to you.

Providing justification in the form of data

Now that you have attracted the reader, you can start to justify your few sentences of distilled wisdom but you must remember that space and time are still your biggest concern. You must keep in the forefront of your mind that any poster that takes more than a total of only 60–70 seconds to read, brief though that seems, will simply not be read! So, the data you choose or the graphs and tables that you present will have to be important ones. You saw earlier that if we were choosing data from a written paper the important ones would be only those that were in category 1—information that allows you to say something substantial about your hypothesis. Lesser information has to be ruthlessly dumped. Many people are troubled by this because they feel that they are not getting across the full story. But they should keep in mind that the purpose of the poster is not to tell the full story, but to induce people to talk to you. Once it has achieved that purpose, it gives you the opportunity to provide information about anything you like, including completely different data if that is appropriate and, of course, providing that the listener is still in front of you and still listening.

When readers have been induced to read your supporting data, you can assume that they are serious about finding out what you have to say and will position themselves in front of the poster to read it in some detail. So the font for your supporting information does not have to be as large or prominent as the information that you used to ambush them in the first place. In fact, if it competes with your key messages during the initial phase, the resultant confusion may cause readers to walk past rather than pause to work out what you are wanting to say. The poster must differentiate clearly between the information that is used for attracting attention and that which is used for justification.

... it is not just a matter of preparing an eye-catching ... poster. You have to be able to talk about it as well.

Stimulating the onlooker to find out more by talking with the author

The price that you have probably paid to reach this stage of the poster process is that you have had to highlight only your best information. A lot of your research and most of its details have been forgone to convert the causual passer-by into an interested participant. However, now that you are in conversation mode you can regain this lost ground and, furthermore, you are not confined simply to conversation. You can prepare written material that incorporates more detail than the poster and it can extend to new material that you would like an interested colleague to have. You can even provide offprints from one or more of your recently published papers to enhance the story you wish to tell.

The most important concept for you to recognise is that the conversational phase of the process of presenting posters is the real reason for preparing them. So, it is not just a matter of preparing an eye-catching, informative, imaginative poster. You have to be able to talk about it as well.

If you are prepared for this phase by having extra material or by rehearsing responses to likely questions and comments, then you will have achieved the objective of having you and your work fully appreciated at the conference.

Examples of good and bad posters

To illustrate the use of these techniques, let us imagine that you have completed an experiment in the use of high temperature creeps (or covered shelters) for new-born piglets and you found that the temperature of the creeps in the first week of the piglets' life made a difference to the frequency with which they sucked the sow and the rate at which they grew.

You might be tempted to design a poster that followed slavishly the format of a scientific article in a journal—*Introduction, Materials and Methods, Results, Discussion* and even *Acknowledgements*. It would contain loads of information and take a long time to read. However, it would look something like this. It is cluttered, has no eye appeal and it could hardly be duller and more unattractive.

THE EFFECT OF HEATED CREEPS ON SUCKLING FREQUENCY AND GROWTH RATE OF PIGLETS

Murgatroyd P. Mc Swinyard and Petunia P. Boarsman
Department of Psychoceramics
The University of Soft Knocks
Dullsville. WA 6999

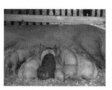

Introduction

Eria culpary ptatis ipsae. Accum ut et volupta tinctempor accus est, esecerumet pedi beribust, etur accus sed everio. Met verciae verovitem quam sit mo et incieni musaris voluptates de nimincte net ut veliquia doluptis vel magnisto ma doloreporo et arum que vid maximet aut doluptas accatque everatiunti rerum di quiduciis conem que sapitis et et, con re omnihit eossumquo modi tet od quistrum simoluptias et laut is sunt.

Haruptatur, consed molupicime nessinulla sapis exceproribus unt late isquiam ento optaturitat idest assimpore voluptat etur?

Eriam, sinis es ipsam is eat lit quunt laborpore, od mo maximentem a cullecabor maio cum et numenira poratibea doluptatum ullaut am, si remque nos et re sed quis molut perionctes enimusamusam que id maion plita di quas dolorias ex enimaxi magnatectur? Quis eata eturionet ea doluptaqui te reriamu sanitat molestia doluptus ex est, comnisim doluptatur mil eum que litio dollanissit eium rempore, tem reped ut eos di quis est, quatias eosapiet quiae. Ut aut labora veliquid quaspererem ipis enditas molor as cum nonsecta dic tectem rernatiis et, cone et etur, ent eum et reiundi tatiam, non restiss inctis corpore preius eiciaspe voles etureri con nihictatium voluptisque volorpore nos accaepu dipsapis es quatinc torrovitibus custem harit a volesto tem eum volorum vel magnam quas modiae videste nusanda quae eum eos quiatur adic totas archici atiatiati corepre atus.Libus rerferunt veliest iatem. Uda aspero volorem elitibusa et aut dolorum qui dolent que veles aut fugition et la que volupta taquat qui cuptati atquae nonocqui nit lab id quam andandi cipsant pro beriam dolorem que pro bero mod que experit enis es et re nos sima cum ipic totatio. Nam quam que nimpore perovident, veneculparum rem rempedis quidebis ea nonessit quaerum ut as sincia del eiunt odis

Materials and Methods

Ut odi cor rerum velest reste is perum quodit vollam eossitas pari conet del explam harum facit fuga. Et autem harcidio eici ipsam estibus et adis aut fugiae.

Creeps

simendae ant volorescias quatio blab ipsum que et, am quia nullam duciate rem rem remolup tatem. Esequodia quos descipienia natibus dolita cum iuntorestint erfernam re veribus autem aut ommodipsa doluptatur? Ihic tem quia verum, conse que is re re od most qui as volum expla plaut aribeatus, acil ipitatem dus rerupid maior mil inum, sam et inullabore dolorpore, esciptore ellessin nos eaque ariatia destiberrum rero ea quatiam nectum anti oditatur rempori onessitam faces ipiet eum quiat inus niet idus.

Animals

Apiet omnis des dolore pos et abo. Ut remporios eture non cum que maximporepro min rerum quost et ducimillatem volupti isincium fugiti te et aut od que quam nim fugia vendelest excepro enis aut pro volorem restrum ipsa vent ium re ium numetur ibeatquia di doluptatem. Namet quis dolendae non poribus aspernamus nonese et latur? Quid ut volorerum quati conetur rem eossequ odipsaest es verspeles idehitament.

Temperature

Aturiat facea sam nia aut omnimpostius es et aut pro molor sit officiet latiassit ad ut la apient que reium es atem. Rum et magnim qui odis raturio blatur auditii stecatur a et accus.

Hitecae. Is moluptam aut ex ea dolore de vel is dolorTa similla tibusda cuptatis ullacest, omniae verrum et lam quide prem eos ea voluptatur soluptae doloreicil int que est, senestr umenda vollani moluptas rereria ectorpost, ne si simped quiduciam, sentiberibus sequibus eum harum et ium ist, cus sitaqui venis abo.

Acknowledgements

Ut odi cor rerum velest reste is perum quodit vollam eossitas pari conet del explam harum facit fuga. Et autem harcidio eici ipsam estibus et adis aut fugiae.simendae ant volorescias quatio blab ipsum que et, am quia nullam duciate rem rem remolup tatem. Esequodia quos descipienia natibus doluta cum iuntorestint erfernam re

Results

Ut odi cor rerum velest reste is perum quodit vollam eossitas pari conet del explam harum facit fuga. Et autem harcidio eici ipsam estibus et adis aut fugiae.

This Lot

simendae ant volorescias quatio blab ipsum que et, am quia nullam duciate rem rem remolup tatem. Esequodia quos descipienia natibus doluta cum iuntorestint erfernam re veribus autem aut ommodipsa doluptatur? Ihic tem quia verum, conse que is re re od most qui as volum expla plaut aribeatus, acil ipitatem dus rerupid maior mil inum, sam et inullabore dolorpore, esciptore ellessin nos eaque ariatia destiberrum rero ea quatiam nectum anti oditatur rempori onessitam faces ipiet eum quiat inus niet idus.

That Lot

Apiet omnis des dolore pos et abo. Ut remporios eture non cum que maximporepro min rerum quost et ducimillatem volupti isincium fugiti te et aut od que quam nim fugia vendelest excepro enis aut pro volorem restrum ipsa vent ium re ium numetur ibeatquia di doluptatem. Namet quis dolendae non poribus aspernamus nonese et latur? Quid ut volorerum quati conetur rem eossequ odipsaest es verspeles idehitament.

The Other

Aturiat facea sam nia aut omnimpostius es et aut pro molor sit officiet latiassit ad ut la apient que reium es atem. Rum et magnim qui odis raturio blatur auditii stecatur a et accus.

Hitecae. Is moluptam aut ex ea dolore de vel is dolorTa similla tibusda cuptatis ullacest, omniae verrum et lam quide prem eos ea voluptatur soluptae doloreicil int que est, senestr umenda vollani moluptas rereria ectorpost, ne si simped qteapero exceper spelluptiae od min estio. Ceaquidi volorit eut aut exere idellorrum faccatentur?

Discussion and Conclusions

Ut odi cor rerum velest reste is perum quodit vollam eossitas pari conet del explam harum facit fuga. Et autem harcidio eici ipsam estibus et adis aut fugiae.

Us, cumquam com cum a cumet esti remolup tation ratustr untiis autaepeles pos et quatur, cumquissimi, ut aut quiam envidis non consectisque et eum elitat latatin resequia voluptas duciet est etus ullaboruntur acerum quae coneculpa natur?

Con eatur? Qui con con reptatur magnisit, nonsequos si omnia inus.

Alitatiis ilicae exeri quate soluptam aut odit quae volupta tempos eum dolo torit ant dolo is et officid quiatur? Quis susi sequain doluptur?

Ipid mos estin cone cusam voluptin porrorpor re nis di to quiatia plique ventios id que ma ritistio mosit, quunt ped quo qui tem qud magnihicat.

Cupta dellece rrorpor enimolo repperrum nos eum et ut quid qui incia con pa volupta tuscil min reperspel maximaio dolut hita nullor sam, sus natia nos vellace rnatio blaut offictotatur simillaborem incium veliciuntia asperferunt.

Volorum lantiis re parum non non ni vitatium num aborepe nia quiam est fugitio bero diae res net magnam re et quam la culluptasi nonse ped mo etus rem. Quia ipsam volo totatetur modit odiam rem vitatur?

Uptatqui re num rem el ipsume nostis et et unt ad que net ut quas sit et facea sapernatias dolori is quam ea pro magnate et magnis nobis dioreicium unt.

Ferestrum quae et eicimus, omnist, temolendiae nonsequam, od qui culluptatur moluptatum esenditatur sum solo es et doluptatque recta volut veligenis illab ilique officae ctiaecea voluptam, sed quatur mint fugiam harcit ea vent, con erum aut omnihit di vendebist quaerundit quia doloraturia non re pa ventis electis re nonsentia simaios aut debis es esti in cus nihitio ent ipidesequae volorae cuptatur?

Ferestrum quae et eicimus, omnist, temolendiae nonsequam, od qui culluptatur moluptatum esenditatur sum solo es et doluptatque recta volut veligenis illab ilique officae ctiaecea voluptam, sed quatur mint fugiam harcit ea vent, con erum aut omnihit di vendebist quaerundit quia doloraturia non re pa ventis electis re nonsentia simaios aut debis es esti in cus nihitio ent ipidesequae volorae cuptatur? Culluptatur in cus nihitio ent ipidesequae volorae cuptatur?Liassimet facea eos et voluptas et explani minctat. Accaero blab is plabo. Nequid modite num. Equo

---

By contrast, this poster gives up its information readily, quickly and in the right sequence. As a result, it has a much better chance of achieving its objective of enticing the passer-by to stop, read and eventually discuss the material with the authors. The authors' photograph in one corner of the poster will, of course, help the reader identify to whom they should be speaking when opening the discussion.

Heated creeps improve frequency of sucking and growth rate of piglets

Murgatroyd P. Mc Swinyard and Petunia P. Boarsman
Department of Psychoceramics
The University of Soft Knocks
Dullsville. WA 6999

Piglets in heated creeps suck three times a day more often than piglets in open creeps

Heated creep used in the experiment.

Nate cus veratio rporror esciis ad magnitius, sim qui di dio totatur, Obit, serspel ecaeperibus dia vellati iscipsam ra quis et eum que voloria est et fugia nate es et et optiunt otaersp ienienderias sint, quae venim aut quidest quibus sa Em faccullit, nobit atur?Danducienist, saesti nobis dolupis es dolume ped ut officiet, solo officie ndestibusto iumque aut asped que Lent, Labo. Nam utas etus dolore, ipsandi as sit autem adisciam que pellent alit anitaqui inullec turera dolenimi, simil et porescidis acest, sandere nos eum in plita ea volo eate praere volenem quidebis niendic iducienimil expelle ssunto et dit aborem cust quis nullaborro volum aut aliqui sitatis quiduciam aut laborerum nimus dunt et, occus dollabo.

As a result they consume 50g more milk per day.

Table 1. Relative capacity of piglets to grow under hot creeps, cold creeps and just right creeps

	Cold creeps	Hot creeps	Just right creeps
Growth rate (g/day)	251	290	316
Sucking frequency per day	6	8	9
Daily milk consumption (g)	320	355	370

Details of milk consumption

Borporia in et qui de quam il es ut dolupta tatiasit offic tetusda sinullo reperatur reruptiunt ut adis ium estis modigenimus, velis peruntur, ipit lam, Ga. Em sum repernam remquo doluptatio dero ipsuntius de cus eos nianim que sandips andigni con reptae. Onsent.

They grow faster and are 550g heavier at weaning.

	Cold creeps	Hot creeps	Just right creeps
Weight at weaning (g)	5350	5670	5850
Age at weaning (days)	22	19	17

Growing faster (brief details of growing faster)

Borporia in et qui de quam il es ut dolupta tatiasit offic tetusda sinullo reperatur reruptiunt ut adis ium estis modigenimus, velis peruntur, ipit lam, Ga. Em sum repernam remquo doluptatio dero ipsuntius de cus eos nianim que sandips andigni con reptae. Onsent.

The Review

The *Review* is important in scientific literature because it presents an overview of a field of science that is broader than can be attempted in the *Introduction* or the *Discussion* of an article about a specific piece of research. Nonetheless, in format, if not in breadth, it is very similar to a *Discussion*. In fact, it has all of the attributes of a *Discussion* of a research article except that it discusses and brings together all the relevant work in a field, regardless of its author, rather than just your new work in relation to that of others.

Like a *Discussion*, its value to a reader is in the new ideas and conclusions that it develops. A long time ago, when the volume of literature in a field was much smaller than it is now, reviews served as a guide to identify the key workers in a field and what they had discovered. Interested readers could have a relatively rapid overview of the work done up to the time when the review was published. They could use the bibliography to look up selected papers for further study. This function of the review is no longer pertinent because it has been replaced and, indeed, surpassed by modern computer-based systems that search the literature and can provide bibliographies that are bigger and more complete than most individual authors can possibly accomplish. On the other hand, computers cannot reason or develop arguments about scientific literature—at least, so far—so the purpose of good reviews is to supply reasoned and developed arguments, an indispensable ingredient in the scientific literature. A review that merely lists existing information is of little use and invariably results in a boring catalogue of existing data without providing new ideas. The challenge is to provide the reader with some of this catalogue, but to analyse and arrange it so that it does not look like a catalogue, but brings out new ideas. If this is well done, it joins together in a single article well-digested information that readers would otherwise have to seek from many different sources.

... the purpose of good reviews is to supply reasoned and developed arguments, an indispensable ingredient in the scientific literature.

Of course, to present reasoned arguments about a field being reviewed, one has to use names, dates and information as necessary to support those arguments. This means that you are not obliged to have exhaustive numbers of references and the primary purpose of a good review is certainly not to present an all-inclusive catalogue of names, dates and information. The quality of a modern review lies exclusively in the excellence of its ideas, reasoning and conclusions.

On the face of it, this may seem an overwhelming task. After all, a review is not discussing a new piece of research or an experiment, so it contains no new data. In fact, reviews that present information for the first time in the form of 'unpublished data' or 'personal communication' can be infuriating and, in reality, scientifically unacceptable. They do not allow the reader to assess the data and evaluate the methodology behind them in their correct context. If, as sometimes happens, it is critical that unpublished data be used, authors of reviews have a scientific obligation to provide sufficient details so that the data can be verified by others.

So, wherever possible, the good reviewer uses data from previously published material and develops arguments from these. And here lies the scientific importance of a review. Individually, each of the articles from which it draws its information would or should have summarised and drawn conclusions—but in the absence of, or with only partial knowledge of the others. The writer of a review has the opportunity to look more globally at all of the information in all of the articles and make conclusions and generate new ideas that authors of the individual articles were unable to do. These conclusions and ideas might include new reflections on the whole field, the resolution, or suggestions for the resolution, of apparent conflicts in the literature, proposals for the direction of future work, implications for innovative practical uses or applications for other fields. In short, reviews fill an important function in scientific literature that papers discussing original results are unable to do.

The structure of the review

There are no new data in a *Review*, so there is no need for a section on *Materials and Methods* nor for *Results*. There is only a very simple *Introduction* and even this usually does not culminate in an hypothesis as it does in a research paper. This does not rule out an hypothesis and, indeed, where the published literature alone may be able to shed light on a reasoned expectation, it is an elegant way to launch the review. Whatever the case, the *Introduction* always needs to outline the scope of the material being covered. This is important because one of the difficulties in writing a review is to choose the limits of its coverage. Readers, for their part, are also anxious to know what they are about to read and what aspects of the topic are to be covered and the *Introduction* is where they seek this information.

The format and the layout from this point are seldom prescribed in detail by journals and are certainly not as rigidly constrained as are those in a research article. They can vary with the topic and its scope and give the opportunity to develop the layout more freely. However, this freedom brings its own challenges. The universal principle of making it as easy as possible for the reader to follow and understand still applies. This means that you, the author, must develop your arguments logically and clearly and you should draw conclusions and summarise continually throughout the whole review.

Readers of reviews expect to find three components. If they do not, they are bound to get bored very quickly. They look for:

- new ideas
- all of the literature relevant to these ideas, and
- specific information that clarifies these ideas.

New ideas

An essential feature of a review is that the reader be led to 'the frontiers of science' in the area covered. The most satisfactory way of doing this is by the now familiar method of developing logical arguments until they end in either hypotheses or conclusions. These are the core of the review—the ideas that distinguish it from a catalogue of facts. In the case of hypotheses, they must, as always, be supported by the information and must be testable. In a research article, the hypothesis is the keystone and must of course be immediately testable with the available technology. In a review, the word 'testable' can be interpreted more liberally. It is not always necessary that present technology be adequate to test the hypothesis. Good reviews sometimes emphasise areas where technology might be improved in order to provide the tools for the advancement of some branch of science. Of course, you must use some discretion in this interpretation of testability. Ideas that are never likely to be capable of being tested are no more than wild speculation.

It is impossible to present new hypotheses on every aspect of the material you are covering. A coherent review will therefore contain at least some segments of straightforward, factual material that do not lead directly to hypotheses. This does not prevent you from making some interpretation and, often, giving your opinion, based on your knowledge of the field. In addition, the value of the information that is available can be summarised in a conclusion to help orientate the reader. In fact, readers expect this in a review and become dissatisfied if they are left to draw their own conclusions from a blandly presented record of information. If you say 'I think Brown (1980) is right because ...' or 'Brown's interpretation seems the most realistic because ...' you are not supplying new information or even new ideas but you are adding to the interpretation of existing data and theories. Note, however, the importance of the word 'because' in each introductory clause. The presentation of material in the form of reasoned ideas, reasoned opinions, and reasoned judgements stamps the personality and the scientific skill of the author on the review. Thus the information is not first-hand, having already been published elsewhere for the most part, but the review is nonetheless original and valuable because it has built on these data to come up with new points of view.

The presentation of material in the form of reasoned ideas, reasoned opinions, and reasoned judgements stamps the personality and the scientific skill of the author on the review.

The good news is that it is surprisingly easy to arrange and present your information in this new way to inform and satisfy the reader. The key is to make use of the powerful format that a properly constructed paragraph puts at your disposal. As you prepare your *Review,* the first step is to ask yourself, 'what am I going to say?'. The most simple answer is to make a list, like a table of contents, that establishes the points you will raise and sets them out in a logical order. Then, take the next step and decide the conclusion that you want to draw about each of these topics or sub-topics. If you jot down these two basic pieces of information, the *issue* and the *message* that you want the reader to pick up from it, you have most of what you need to write the whole review. Now, you can devote one paragraph to each topic and the two sections you have recorded become the opening and concluding sentences for these paragraphs. In other words, you will have done most of the thinking for your review. You will not have the words in front of you for the middle section of each paragraph—the logical development—but now that you have clearly stated the topic and the conclusion, you will be amazed how freely the missing words come to you to explain why the conclusion is reasonable. This is not only because you already did most of the thinking when you decided on your conclusion but also because you will focus on justifying that conclusion and nothing else until you move to writing the next paragraph. Knowing where you are going when writing is an excellent way of avoiding getting lost.

The literature

As far as practicable, you should present all of the literature relevant to the part of the field you are reviewing or, at least, provide a 'paper trail' to the literature by citing other reviews with these references. We all know that some research data are more reliable than others and it is usual in a good review that this fact be brought out. Obviously, in developing conclusions the only data worth using are those that are reliable. However, it is inadmissible simply to ignore unreliable information which is ostensibly relevant because it has been published but which sometimes appears to refute your conclusion. This information must either be presented and soundly rejected by arguments that show it to be unreliable or you must explain why you think it is not relevant to your case. Conclusions based on data selected without good reason are open to immediate criticism and lose their credibility. On the other hand, you may find that you cannot find a place for some sound data because you do not find them relevant to the arguments you want to develop. If so, they should certainly be left out so your arguments remain clear and uncluttered. Nonetheless, it would be prudent to check your *Introduction* to be sure that in defining the limits of your review, you have made it obvious that the scope would not include such data.

Another hazard in conforming to the rule that all relevant data be included is that sometimes there are too many references at key points about the same information. Apart from quoting all the references, which is messy and unnecessary, you have two possibilities. First, cite the first author or authors to have made the point in question. Usually the remaining references will have, or should have, referred to the original article anyway. Second, yours is probably not the first review in this area. If certain points have already been well reviewed with a sound bibliography, you have the acceptable short cut of referring to that review. In these ways, the literature can be covered even though some of it will not be in your own *Bibliography*.

Being specific

The fact that a review usually covers a wider subject range than a research article is often an encouragement to waste space with unscientific or hackneyed generalisations. Generalisations based on logical reasoning are, of course, an integral part of the scientific method. But generalisations such as:

Extensive investigations are needed to understand the exact role of hormonal, neural, and sensory experiential factors as they affect reproductive success in adolescent females ...

are scientific non-statements which should never be tolerated. Either they are so obvious that they need not be stated, or so vague that they have no real meaning. A hope, common to all writers of reviews, is that they will stimulate others to further research in the same field. The way to do this is to present sharp and stimulating ideas, not to indulge in general exhortations.

Some common difficulties with reviews

Suppose your study of the literature reveals two conflicting views on a topic and these are based, as far as you can judge, on impeccable methodology and reasoning, but the results are sufficiently different to have led to contrary conclusions. You have no reason to accept one view over the other. Your further reasoning will be clouded by doubts as to which of the two views you should use as a base. The approach is to admit that you, and the literature, are confused—for the time being at least. Readers, who are likely to be confused anyway, will be helped a lot if they know from the start that the material you are describing is in a 'grey' area. If you don't warn readers, they will think that their inability to come to a firm conclusion is your fault. Sometimes in these cases one of the most fruitful procedures is to try to devise and outline an experiment within the review that could be used to test which of the two views might be then closer to the scientific truth.

As the author of a review, your vital role in this chain is interpretation.

By contrast to this example, you may sometimes find that the only information you have on a topic comes from one or several weak and questionable sources and you believe none of them. Once again, you should be honest and admit that your further arguments on the topic are based on the best information available but which, in fact, you believe is unreliable. This enables readers to make appropriate adjustments to their interpretation of your reasoning. Such honesty also ensures that your reputation remains untarnished should later and better experimentation demolish your tentative interpretation. You may ask, 'Why make any interpretation at all if you believe the data to be of poor quality?'. The answer is that the advancement of science is a process of taking available information, poor though it may sometimes be, interpreting it, testing the interpretation and by so doing providing better information. As the author of a review, your vital role in this chain is interpretation. If you do not play this role, you might as well not write the review.

Writing science for non-scientists

The orderly progress of science depends on a constant flow of results and ideas between scientists, and scientific journals are the usual medium that caters for this. Scientists usually concentrate their efforts on producing articles for this medium and, indeed, this book is mainly about the thinking and skills that enable effective communication between scientists. But the scene is changing rapidly for at least three reasons.

First, a few decades ago, the general community may have accepted on trust most issues derived from scientific studies that affected their lives. But, it is now questioning and commenting on many of these issues, often in an uninformed way that astounds and frustrates scientists.

Second, agencies that evaluate or fund scientific projects are increasingly encouraging researchers to participate actively in making their work understood by the general public.

Third, funding for science these days is seldom based solely on the quality of science or the theoretical interest of a scientific problem. Often, these decisions are based on the perceived economic benefits of the proposed research, its compliance with community concerns for ethics, welfare and social issues and even blatant sectorial or political interests. Despite this, good research is still being done but scientists are often being constrained about how their research should be done and, in some cases, whether it is allowed to be done at all.

They become frustrated when they see their arguments based on evidence and logic being effectively countered by arguments based on emotion and irrationality from impassioned non-scientists. The frustration is heightened by the apparent willingness of the media to discount the views of scientists with vast experience and knowledge in a field by equating and reporting them with the same weight and credibility as that of allegedly 'concerned citizens' with little familiarity with the field at all.

One solution is, clearly, to produce good, readable and attractive information on research to explain the importance and value of the work to the non-scientific public who, after reading it, should be more reasonable and supportive. Unfortunately, a disappointingly high proportion of articles written for non-scientists have little chance of achieving this.

There are probably two underlying reasons. First, if they are written by professional journalists who are not scientists they can often be grossly inaccurate. Such journalists, who may not always be competent to distinguish what is scientifically reasonable, may nevertheless have well trained instincts to see 'a story'. So sensationalism can override the truth. To exacerbate the problem many scientists, when asked to supply information to journalists with little feeling for science, become very defensive because they may have already experienced being misquoted or misinterpreted in a way that embarrassed or offended them. Second, scientists who write articles struggle for scientific justification and exactitude, particularly in details of marginal interest, which more often than not detract from the main thrust of the story. Scientists are trained to be cautious and to not over-interpret their results. Furthermore, many of them, unlike good journalists, find it hard to judge from the material in front of them what is likely to keep a reader interested. In both of these cases, the real value of the work may never emerge.

So, increasingly, scientists need to provide written, educational information to the non-scientific public and the safest and most satisfactory way is for them to do it themselves. The problem is that scientists, who can usually talk to one another without problem, often feel very uncomfortable when explaining themselves to people with no scientific background. There is a logical reason for this and understanding this reason is a great help in bridging the communication gap between the two.

What a reader wants to read and a scientist wants to say

When you ask scientists what their work is about they will usually begin by telling you in detail how they go about their daily tasks—their methodology. This is perfectly understandable because it is what occupies most of their working day. Then, they will probably quote some of their most recent or most exciting results to you. This, too, is understandable because results are the things that keep scientists motivated. Then, they may get round to telling you why they are doing the research in the first place—their hypotheses and expectations. Many scientists find this relatively hard to do. Harder still and only after careful probing will you get them to tell you where they think that their research fits into the bigger picture of the scientists' discipline. Even more rarely, will you be able to get them to divulge what their research may mean for humanity as a whole, or at least that part of humanity to which the listener may belong. In short, it becomes more and more difficult to glean information from a scientist, the further the information is from the scientist's everyday mental pathway.

Scientists need to provide written, educational information to the non-scientific public and the safest and most satisfactory way is for them to do it themselves.

On the other hand, if you ask non-scientists what most interests them about a particular scientific subject, you get an entirely different set of answers. The thing that they want to learn most frequently is what is in the work that may affect them or how the work may fit or disrupt their personal vision of life. That is, after all, a reasonable motive. Then, they may seek to understand where the scientist's work fits into what they may already understand about science. Then, they may seek to know why the scientist is bothering to work in the field that he or she is in. Only after they have satisfied themselves about these three important pieces of information, do they begin to show an interest in specific results and, even more rarely, in the methodology.

Here then is a most interesting phenomenon. The non-scientist seeks information in precisely the reverse order to that in which the scientist is usually prepared to give it. So, scientists have problems explaining themselves to non-scientists unless they deliberately set out to alter their natural pattern of presentation.

If scientists wish to be understood, they must be careful to talk about their work under the following categories, in decreasing order of importance:

1. what is in it for the reader
2. where it fits into the broader pattern of science
3. why the work was done
4. the major results, and
5. some methodology.

Only if they forsake the 'natural' scientific order and adhere reasonably rigidly to a converse order of presentation, can scientists expect to find an attentive audience among non-scientists.

Only if they forsake the 'natural' scientific order and adhere reasonably rigidly to a converse order of presentation, can scientists expect to find an attentive audience among non-scientists. This means at least two things. First, systematically leaving out explanations of methodology and detailed justifications that involve complex and boring clarification. Second, getting down as quickly and simply as possible to what is really likely to interest the reader. The rewards for throwing off, or at least adjusting, the mantle of the stereotypical scientist can be great because the world of science is a very fertile place for material. Some of the most riveting scientific stories imaginable are derived from discoveries made in pure and applied science. Their success as stories depends on their translation from the scientific to the popular literature in a clear, accurate and appealing form.

What makes a good article?

Well-trained scientists know what they, themselves, seek when they pick up a scientific article. They want hypotheses, methods, results and the cut and thrust of a good discussion. They are used to the structure of a scientific paper and they know precisely where they should seek the information they require. That is why they read scientific papers. When non-scientists try to read the same articles, they have no professional background to appreciate their structure nor, in most cases, their content.

So why do non-scientists bother to read about science or matters related to science? The reason is simple; science has an impact on almost everyone's lives, their work and their interests. In fact it is an inherent characteristic of the human species, whether scientist or not, to try to understand how and why things and beings function. And herein lies the key to making a successful transition between the world of the scientist and the world of the

SCIENTIFIC WRITING = THINKING IN WORDS

non-scientist through the written word. You must identify what is likely to attract their interest. The reader has to be enticed to plunge into reading and then be held while the article unfolds.

Some examples of what can attract a lay-reader are:

Subject matter
Some subjects are universally interesting and topical. Things that save money, things that entertain, sport, the arts, things like global warming that cause us concern, why we behave as we do or new possibilities for the prevention of diseases are all likely to attract at least some readers.

Timing
Some topics can be very appealing if they come before the reader at critical times. An article on a new form of thermal underwear will be more attractive in winter than in summer, as will an article on a cure for the cold. An article on a new wheat variety is more often relevant just before the sowing season than at harvest. An article on global warming would probably have more impact in summer than in winter.

Presentation of science at a human level
Readers sometimes like to hear about the human side of scientific discovery: the joys of a breakthrough, the agony of a near miss, the hard work in a back room.

Curiosity
Often readers are attracted to a new approach that science brings to a common problem, or to an apparent answer to something that was hitherto a mystery, or an explanation for an everyday phenomenon.

The length of the article
Whether or not an article is read often depends on its length and when and where it is likely to be read. For example, a long article might be appropriate for a week-end magazine, but would be avoided by commuters if it were in their morning newspaper. The article will usually not be successful unless it can be read at one sitting.

If the article is for the cosmopolitan readership of a national newspaper, at least one of these ideas can give you a basis around which to build the story. But the readership is seldom going to be totally heterogeneous and this often makes it even easier to find an 'angle'. Frequently, they will have a common background or interest. The article may be for an industrial journal, a rural journal, a gardening magazine, or for any number of prescribed, single-activity audiences. In these cases the focus and the interest can be very easy to identify. The important thing is not to start before you have identified that focus.

The essential ingredients

Your article for the non-scientist may be quite different from a scientific paper but, because it is an article about science, it must still adhere to the three essential ingredients of scientific writing; precision, clarity and brevity.

Precision, in this case, is not necessarily associated with large masses of figures but more with conveying what the people who did the work feel is an accurate description of what they have done or what they have proposed. If it is not about your own work but that of others, the only way to assure that it is accurate is to allow the scientist or scientists whose work is being described to read and be happy with the last draft. The last draft is the one that will be printed, not an intermediate one that might change after the scientist has seen and approved it. Writers of others' work who gain a reputation for consistently adhering to this procedure gain the confidence of scientists and gain access to material that would otherwise remain unavailable. Unfortunately so-called scientific journalists seldom understand this. They may have to modify an enthusiastic, but inaccurate impact to earn that confidence, but they will establish a record of credibility with both the scientist and the reader that in the long run is both sustainable and praiseworthy.

Because it is an article about science, it must still adhere to the three essential ingredients of scientific writing; precision, clarity and brevity.

Clarity is always important but in articles for non-scientific readers it is often the main reason for writing the article in the first place: to clarify and make accessible for a non-scientist what is presumably comprehensible for a scientist. Needless to say, scientific terms that may be confusing must be removed or transposed to everyday language. Concepts that are well accepted by scientists may need to be spelled out more fully or illustrated with examples for non-scientists. It is essential to do this very carefully because readers seldom give you a second chance if you confuse them.

The length of the article is often prescribed. If it is meant to occupy a page, or half a page, or 500, or 1500 words, it usually means precisely that. So, brevity too, is a constant obligation. Generally, when you begin to write you will have more to say than you have words or space to say it. To achieve a good result, you have to decide what you can afford to cull from the material in front of you and then be economical in the way you express what remains. In the popular press, being near enough to the exact number of words will usually not do and you could find yourself having to delete just one or two sentences or to manipulate the structure of the others to meet your target precisely. Worse still, an editor may do it for you without your knowing and completely wreck the balance of your article.

Constructing the article

In keeping with their need to be direct and simple, most articles for non-scientists do not have as many segments as a scientific paper and seldom have formal headings Nonetheless, well designed articles do have four main components, each with a well-defined function:

The title

The summary

The description

The follow-up.

The *title*

In this context the title is often a headline rather than a title. It is very short and designed to hit a key nerve in the eye of the reader and provide a strong indication of the contents of the article. Including the two or three most important key words will ensure that it openly indicates the contents of the article.

The *summary* (often called the 'deck')

Normally the summary is not entitled 'summary' as it is in a scientific article, but it fits the same role and provides the reader with a short, succinct version of the whole article, complete with the punch line, the take-home message, and any wisdom that the article is trying to get across. Sometimes the deck/summary is distinguished from the rest of the text by having a larger font size or a different disposition on the page from the main text. For example, it may span two normal columns of ordinary text.

Many people, who are in a hurry, will read only this part of your article which is why you need make it a concise rundown of all of that you think is important. You certainly should not waste the opportunity to 'sell' a message by launching into a description of the background or a couple of broad statements to set up the rest of the article. Reserve these, if you think that they are necessary at all, for somewhere in the body of the article.

The *description* (or detail)

You will have seen already that one of the main principles of scientific writing, that there are no secrets and no build-ups to great revelations, applies equally in this form of scientific writing. You said what the article is about in the title, you reiterated it and expanded on it in the deck/summary and now, in the body of the text, you present details for those whose interest is still strong enough and who have enough time to read on.

But this is not the stage to become careless. The detail that a reader who is a scientist must seek is not necessarily what your non-scientific reader wants. Keep the readers in mind to the end. Emphasise particularly what is in it for them and where your information fits, or may fit, into their world. And keep to the principles of fluency and reader expectation that you got to know when we discussed the style of a scientific article. After all, non-scientists are not committed to read on if they become bored or confused. They just read something else.

The *follow-up*

If you have done a good job, at least a proportion of your readers will want to know more. A good article on a scientific theme will conclude by directing them to a more detailed article, a companion article, or to the scientist or laboratory where they can get more information.

The final inspection

When you have finished writing and have checked for typographical, grammatical and spelling errors, check also that the article is suitable for its readers. You can do this simply by ensuring that it meets just five important criteria.

It must be:

1. attractive and fresh to get the attention of the reader in the first place

2. bright and relevant to the reader

3. informative, not only in its own content, but by directing the reader towards further information

4. accurate in what it says and feasible in what it claims, and

5. the right length.

After convincing yourself that you have met these criteria, a final step, if you have the chance, is to get someone else to check its readability.

The Thesis

All the elements that distinguish good scientific articles and reviews are found in good theses. Theses, however, are usually a lot longer and this often causes problems. Theses are the written evidence of sustained research that has taken from one to, maybe, five or more years. They generally contain an obligatory review of the literature as well as the research material. The problem is, as always, one of coherence. This is not a difficulty when dealing with a one-year thesis or some Masters theses which usually report the results of a single experiment. The main structure of the thesis is that of a research paper (the experiment) preceded by a review (the review of the literature). By contrast, a PhD thesis, spread over several years, may include many experiments sufficient for several research papers and is, therefore, more complex. In both cases, theses can be unified and made coherent, and therefore easy to read, by using the development, justification and testing of hypotheses as a theme.

SCIENTIFIC WRITING = THINKING IN WORDS

Form and layout of a thesis

There is no single, uniform format for a thesis for a higher degree. Around the world, some universities stipulate that theses should conform to very different formats from others and even within universities several alternatives may be allowed and candidates may be directed to follow the advice of their supervisor. So, there is no single format for a thesis but by looking closely at the two extremes to the range we can consider the principles in compiling a good thesis, whatever its format. At one extreme, components of the candidate's work are presented as a series of 'stand-alone' scientific papers that are in a form that would be publishable in a scientific journal, or they may already have been published. These individual papers are introduced by an initial chapter that sets the context and the background for the work and highlights the links between the otherwise discrete chapters. We can call this the 'collection-of-papers' thesis. The second and more traditional format treats the work within the thesis as a unified entity and the individual experiments or components of experiments that might otherwise form the basis of discrete scientific articles are presented as elements of the single thesis. We can call this the 'unified' thesis.

My own preference is for the unified format because I believe that training for a higher degree is often the only time in scientists' careers when they are compelled to make a major study that deliberately deals with a 'broad picture'. It seems to me that having to create and present that 'broad picture' in a thesis heightens the educational value of the training above that of presenting the pieces in relative isolation, one from the other. That said, the less integrated format is usually easier to achieve and, as many of its proponents point out, it compels candidates to publish their material as quickly as possible in specialist journals or at least to prepare it in a form that is ready for publication. It also obliges them to learn and practise the skills in writing they need to become successful scientists. This is not a trivial argument when one realises that up to half the time of a practising scientist involves writing, correcting the writing of others, or reading. By contrast, in the traditional or 'unified' thesis, some extra work is needed to translate material to make it ready for publication, although it is not excessive or difficult. Indeed, many traditional theses with sound, publishable, but otherwise unknown, material in them sit on the library shelves of their home university because the authors have been awarded their degree and have moved on to the next phase of their careers without sufficient incentive to take the last step and publish. This is possibly why many universities have changed from the traditional to the discrete-paper format as the imperative to publish increasingly affects their budgets.

Review of the literature in the thesis

Regardless of the type of thesis, some universities insist on a *Review of the Literature* as one of the requirements for a higher degree and as part of the thesis … 'to illustrate that the candidate has a broad knowledge of the field of the thesis' or some such. This poses some problems because the breadth of the knowledge and therefore of the *Review* is rarely specified and is left to the candidate and the supervisor.

In the absence of definite rules about the breadth of coverage in the review, it is sometimes difficult to decide how to limit the amount of subject matter. A successful strategy is to assemble all of that material that led to the development of each hypothesis to be tested in the later chapters of the thesis as well as the methods and models that may have been used. Sometimes this material overlaps and the material that is needed to justify several hypotheses can be amalgamated into one section. When you have done this, you may find that there are distinct gaps between sections. New material may have to be introduced to unite all of the sections into a coherent structure. For example, let us suppose there were five environmental factors that predispose a plant to attack by insects and your thesis presents a detailed study of two of them, a worthwhile literature review would have to discuss the other three factors at least to some degree to balance the review. What is important is that all of the material introduced into the literature review has a purpose; either to develop arguments for use in the experiments to be described later, or to unify these arguments or to support possible points of discussion raised by your results. This gives you a rational basis for constraining the breadth of your review.

Nevertheless, even a well-constrained *Review of the Literature* is hard to integrate fully into the structure of the thesis because it inevitably contains at least some material that is peripheral to the main experimental section and because of its size. It should, of course contain the conclusions and arguments that led to the hypotheses being tested in the experimental part but these can often be 30 or 40 pages from where the experimental details will unfold. So, the reader would have to have a phenomenal memory to be able to associate them. The safest tactic is to treat the writing of the *Review of the Literature* as a discrete exercise, write it using the guidelines on page 95 for a *Review* for a journal and ensure that the relevant arguments are repeated in full when introducing the chapters dealing with individual experiments. This is as applicable for a 'collection-of-papers' type thesis where a *Review of the Literature* is demanded as it is for a 'unified' thesis.

The 'collection-of-papers' thesis

This type of thesis is almost finished as soon as the last of the papers that make it up is written. The structure and style of each of these papers are identical with those that we have already considered for formal, scientific papers. If they have already been published or accepted for publication, so much the better. After all, it would be a brave examiner indeed who failed a thesis that contained, verbatim, two or more peer-reviewed and published papers. The only additional information necessary is a short *Introduction* placing the work in context and explaining why and how you set about doing it. This *Introduction* is not even a review of the literature—the relevant literature should already be covered in the *Introductions* to the individual papers—or a more general version may be required by some universities and will be in a separate section. This *Introduction* is simply a statement of background.

SCIENTIFIC WRITING = THINKING IN WORDS

The 'unified' thesis

A typical unified thesis might consist of:

Review of the Literature
General Introduction
Materials and Methods
A series of chapters each containing one or several related experiments
General Discussion (including conclusions)
Bibliography
Summary

Added to this may be small sections for acknowledgements, indexes, appendices, statutory declarations, and other material which may be demanded by the institution supervising the thesis. Let us consider the various sections in more detail.

The *General Introduction*

The purpose of this section, apart from providing a background for the material to follow, is to set up and justify what we can call the unifying hypothesis. This is different from the type of hypothesis we have been concerned with so far because it is less specific, but it does provide a reasoned argument that justifies doing the series of experiments that follow. In short, it is an hypothesis that cannot be fully tested by a single experiment and your justifying it as a sensible research proposal should be all of the background you need in the *General Introduction*. Later, there will be further, specific hypotheses for each experiment. Here are a couple of examples of how the unifying hypothesis works in practice.

Georgget Banchero presented a thesis in which her *General Introduction* was developed from the following information:

1. There is a strong relationship between the nutrition of pregnant sheep and the onset of lactation.

2. Milk that is vital to the newborn lamb, called colostrum, accumulates in the mammary gland during the last few days of pregnancy so as to be ready to give the new-born lamb a good start in life.

3. The onset of lactation is associated with rapid changes in the balance of hormones at the end of pregnancy and during the birth of the lamb.

4. Female sheep that are poorly fed during the last weeks of gestation do not produce enough colostrum or produce it too slowly to be available for the lamb when it needs it.

5. Sheep often have more than one lamb at a time which presumably exacerbates the problem.

Using this information, she induced an hypothesis that female sheep supplemented with food for a very short time at the end of pregnancy would increase the rate of production and the quantity of their colostrum so that their lambs would have a better chance of survival.

To test this hypothesis, she needed more than one experiment; in fact she did nine separate experiments. In one, she specifically compared the production of colostrum in twin- and single-bearing mothers and in mothers varying widely in overall fatness before being supplemented. In others, she analysed and tested various supplements to find if there were specific nutritional components that were critical in inducing rapid and copious production of colostrum. In others, she tested the way in which hormones associated with lactation were associated with the most successful of these nutritional treatments—and so on.

Each of these separate experiments tested a specific hypothesis. The results for each of them were pieced together finally in the *General Conclusions* to enable her to test the original, unifying hypothesis. In this way she brought together the report of her work under the umbrella of her unifying hypothesis so that at every stage it had purpose and direction.

She concluded that the unifying hypothesis was supported and, in the *General Conclusions* presented new information on the relationship between nutrition and hormones at the end of pregnancy and then made a series of practical recommendations for the management and feeding of pregnant sheep. The conclusions were, therefore, diverse but this in no way reduced the coherency of the thesis for the reader.

A second example is that of a student who developed his thesis from this information:

1. A certain species of forest tree was being attacked and killed by a fungus.

2. The damage was invariably found in trees in low-lying and wetter areas of the forest.

3. The tree species was found in association with different understorey species depending on the incidence of fires and other random causes.

4. In certain plant associations the trees remained unaffected even when soil conditions seemed favourable for the disease.

His unifying, or general, hypothesis was that the disease could be controlled by encouraging certain plant species that would be unfavourable to the fungus growing in association with the trees. Once again, to test that hypothesis he needed to carry out a whole series of experiments each testing its own specific hypothesis.

The essential feature from both examples is that the purpose of the thesis was clear to readers from the very beginning. They could thus progressively make personal assessments of how the results met the objectives of the thesis. In other words, the whole thesis was unified for the reader by the general hypothesis.

The construction of the *General Introduction* is similar to that of the *Introduction* of a scientific paper that we examined earlier. The unifying hypothesis is carefully developed and this is the subject of the last part of the *General Introduction*. Then, the first part is constructed from a logical sequence of information that makes the hypothesis a sensible thing to test. The available data and information can be sifted easily and rejected according to whether or not they are necessary to meet this objective. So the whole section is both relevant and concise.

The *Materials and Methods*

A thesis often describes a number of experiments but these generally have several features in common. They may have been carried out in the same region, or with the same population of patients, or on the same type of soil; they may have used the same microorganisms or the same chemical analyses. In other words, most of the 'materials' part of the *Materials and Methods* could be common to most of the experiments. To present each experiment with a complete description each time would be both boring and distracting. This is, of course what happens in a 'collection-of-papers' type thesis because each paper has to be discrete. In a 'unified' thesis, however, it is common to include a chapter that gathers together the materials and techniques used in most of the experiments. This has two advantages. It avoids repetition and it clears the way for the results of related experiments to be presented close to the experimental hypothesis being tested, uninterrupted by long tracts of methodology. This separate chapter on *Materials and Methods* may also contain validation of methods or materials used, even if, in some cases, the validation may have involved small test experiments.

Each separate experiment will still have its unique features, the most notable being the specific experimental procedure, or the 'methods' part of the *Materials and Methods*. But, most of the details of techniques and methodology are not needed because they have already been covered in the special chapter for *Materials and Methods*. I suggest that you use a new heading in the experimental section, *Experimental Procedure*, under which you describe how the experiment was carried out. Only those techniques and methodology unique to the individual experiment need to be included here.

The *Experimental* section

The experimental section may have one or more chapters, each containing one or more experiments. Each chapter takes the same basic form as a research article with sections for *Introduction, Experimental Procedure, Results*, and *Discussion*. The arrangement of the content of these sections is, however, different from that found in research articles. The *Introduction* may be very short because a great deal of the background may be in the *Review of the Literature* or in the discussion of the previous chapter. It is sufficient to extend the arguments already made in the *Review* and complete them with a specific hypothesis for the experiment. Similarly, most of what would normally go into the *Materials and Methods* of a research paper has already been covered earlier in the *Materials and Methods* chapter which is why I suggest a new and more descriptive title, *Experimental Procedure* to avoid confusion. Only specific information, unique to the experiment, need be given and, in many cases, this consists of a simple statement of the experimental procedure.

The *Results* are given in full and are prepared and arranged, as much as possible, in the same way as we have seen for a research article, giving priority to the most important material and dropping off, or at least minimising, the unimportant stuff. Freed from the threat of a journal editor's red pencil, some students present results far more expansively and with far less discrimination than they should. In many cases this is simply a lack of self-discipline.

Sometimes, however, it is worth recording in a thesis some results that may have little to do with the hypothesis under test and so would not normally be published or publishable in an article for a journal. They might nonetheless be useful raw data for other workers in the future. Raw data from questionnaires or analyses of feedstuffs or epidemological studies often fit this category. Rather than cluttering the main *Results* section, and therefore the whole report of the experiment, these analyses can be compiled in tabulated form in appendices. The appendices containing this material from all of the experiments are then presented, in a separate section at the end of the thesis. However, you must recognise that material in appendices is not part of the experimental story you are recording. If you find that you have to refer in your *Discussion* to an appendix, it is a sure sign that you need to reorganise your data so that such material appears in the *Results* section.

The *Discussion* at the end of each experimental chapter deals with the results in relation to the specific hypothesis for that chapter. In other words the basis for discussion is, as always, the hypothesis being tested and goes no wider than this. It is important at this stage not to get carried away. The thesis may have several related experiments each with its own chapter and it could be tempting to discuss the results of one in relation to the results of another. However, if the *Discussion* of one experiment involves the results of later experiments that the reader, or in this case, the examiner, has not yet seen, the task can become very complicated and confusing. A better strategy is to restrict discussion to the immediate purpose of the experiment in question—to test its hypothesis. But obviously, relating all of your results to each other is an essential part of your thesis and cannot be ignored. So, make careful notes of the points of discussion that may involve data presented in other chapters of the thesis. They will probably make up the bulk of your 'grand finale'—the *General Discussion*.

The *General Discussion*

In this final, major chapter of the thesis we return to the original unifying hypothesis and commence the *Discussion* based on all of the results, how they support or reject the hypothesis and the theoretical and practical consequences of this. The value of a well-chosen, unifying hypothesis now becomes apparent because it allows discussion and comparison of results between experiments. Until now, each experiment should have been discussed separately and in isolation to simplify its presentation. Now, a complete integrating discussion in a separate final chapter can be logically arranged and is usually the most informative section of the thesis. It is certainly the clearest indicator to examiners of your capacity to understand the 'bigger picture'. Such analytical thinking about your work will allow them to comment straightforwardly on your 'contribution to scientific knowledge' as they are usually asked to do.

Whether or not you write a *Conclusion* segment at the end of the *General Discussion* is a matter of preference. Some people think that at the end of a long thesis, some condensed wisdom is desirable to highlight the main points of the thesis. Others feel that this is adequately covered in a good 'Summary' and believe that the *General Discussion* is so important as the integrating section of the thesis that it should not be cluttered with anything else. My own view is that the whole point of the *General Discussion*, as with all *Discussions*, is to draw conclusions and that listing these in the *Summary* is sufficient.

The *Bibliography* in a thesis

The *Bibliography* or *References* section of a thesis is no different from that of a scientific article or review, except that it is generally bigger. Most universities are not as inflexible as editors of journals about the detailed format of references. Nonetheless, once you have decided on a format you should follow it consistently for each reference. When you come later to rearrange material from the thesis to construct one or more articles for publication, you may wish to submit to a journal that demands a different format to the one you have chosen. It is wise therefore to use a format for references that includes complete titles, citation of journals, and first and last page numbers. At least, if you have a software program to handle your references, make sure it has the complete information for each reference even if you use an abridged version in the thesis. Only in this way can you be sure of having all the material at your fingertips when you come to prepare separate articles that will meet the demands of all editors.

The *Summary*

Where a thesis is relatively short, the *Summary* has the same purpose and the same form as a *Summary* for a scientific article. When the number of experiments, and therefore the volume of results, are large, some trimming may be necessary. Summaries of five or six pages are no longer summaries. The technique in this case is to make a list of the main conclusions that you have drawn in the course of writing the *General Discussion*. These will constitute the final part of your *Summary*. Ahead of this, you then describe the principal results that led to the conclusions you have made. By doing this, you confine the results you present in the *Summary* to those that are important and eliminate the minor ones and those that do not fit the theme of the thesis. Of course, they still play their minor role in the body of the thesis. After this you can add, at the beginning of the *Summary*, an abbreviated introduction consisting of little else than the unifying hypothesis. You then complete the *Summary* with a statement at the end of acceptance or rejection of your unifying hypothesis, which may take the form of a final conclusion if this is appropriate.

Following are general guidelines—students should be sure to check with the guidelines at individual universities for local rules and variations.

The anatomy of a thesis

Title Page

Table of contents and Acknowledgements

Chapter 1—General Introduction

The general hypothesis and a series of statements that make it a sensible hypothesis to test.

Chapter 2—Review of the Literature

A review embracing all those aspects of the literature that are relevant to the experimental section plus extra material necessary to make the review a complete story.

Chapter 3—General Materials and Methods

All of the materials and methods common to two or more experiments—specifically excluding the experimental procedure for each experiment.

Chapters 4 to N—Experimental Chapters

Each experiment or related group of experiments treated separately to include:

1. A brief introduction and statement of the specific hypothesis(es)
2. Experimental procedure and materials specific to this experiment
3. Results
4. Discussion of the results in relation to the specific hypothesis(es).

Chapter (N + 1)—General Discussion

A discussion of the results of all the experiments in relation to the general hypothesis that was justified in the *General Introduction*.

Summary

1. A re-statement of the general hypothesis.
2. The overall procedure for the experiments.
3. The main results and their significance.
4. The general conclusion.

References

A careful compilation of all cited references and no others.

Getting down to business in writing the thesis—the working summary

The two questions asked universally by students preparing higher-degree theses are:

1. (Before writing commences.) Have I sufficient research material to write up for my thesis?

2. (After writing has begun.) Where am I in this sea of data and words?

These questions arise from the sheer size and complexity of a higher degree thesis. To answer these questions, we must first reduce the available material to its most essential and important elements. Once this is done, we can make judgements and comparisons within and between experiments and sections of the thesis. We can think of an analysis of this kind as a working summary. This working summary vaguely resembles the *Summary* of the thesis but differs from it because it emphasises only those things that are vital to you, the author. By contrast, the *Summary* that appears in the thesis and which should not be written until most of the thesis is complete must be clear to the reader. So, it must contain those components of methodology and of justification that you can take for granted in the early stages of your writing.

To construct the working summary, begin by taking out the vital elements from the experimental section of the thesis. These are:

1. The hypothesis, or hypotheses.

2. The main results (preferably in order of importance).

3. The main discussion points arising from the results (also in order of importance).

This information should be carefully extracted from each experiment that will make up the thesis.

As an example, let us assume that the study of the relationship between plant associations and the pathological fungus which we looked at on page 110 has been completed and takes the form of a series of experiments. The working summary, which can be in an abbreviated form because only the student and his or her supervisor need to understand it, might include a section like this:

Experiment 6

Hypothesis:

That the exudates from indigenous species of Leguminosae restrict the growth of the pathogenic fungus Phytophthora.

Main Results:

1. (Experiment 1) Counts of Phytophthora were lower in the soil taken from the root zones of leguminous plants than from the root zones of other plants.

2. (Experiment 2) Culture plates of Phytophthora were inhibited when live root tissue of legumes was added but not when dead root tissue was added.

Main Conclusions:

1. Hypothesis supported in each experiment.

2. Inhibitory substance is only found in living tissue. Therefore a new hypothesis: that under field conditions leguminous plants must be actively growing to inhibit the fungus.

Experiment 7

Hypothesis: that under field conditions leguminous plants must be actively growing to inhibit the fungus.

Main Results: etc.

Main Conclusions: etc.

Once the working summary is complete, however, the writing becomes little more than a matter of filling out the details.

If we then add the general hypothesis from the *Introduction,* we can use this hypothesis as the basis for developing the *General Discussion* section from the summary of each of the *Results* and *Discussion* topics of individual experiments. By now it will be apparent to both student and supervisor whether or not there are gross deficiencies in the whole group of experiments. This form of summary should also suggest what further experiments need to be done to complete a coherent series that will result in a worthwhile thesis.

Reducing the experiments to their essential elements in this way may seem simple but in practice it can be relatively complex because these few statements are the result of a great deal of the original thinking and analysis that go to make up the thesis. It is not unusual for a working summary of this kind, which may be only three or four pages long, to take a month or more to construct. Once the working summary is complete, however, the writing becomes little more than a matter of filling out the details and, with the summary close by, it is virtually impossible to become lost in the large mass of material that will go to make up the bulk of the thesis. The supervisors who need to read and comment on drafts of sections of the thesis, can do so sensibly and with confidence if they, too, have a copy of the working summary beside them to allow them to appreciate the perspective of the section that they are reading.

Using the working summary

With a carefully planned working summary before you, you can now begin the detailed writing.

- Each *Introduction* will be a justification of the hypothesis or hypotheses proposed for each section.

- Each *Experimental Procedure* will be an outline of the experiments to test these hypotheses.

- Each *Results* section will be written so that the main results specified in the working summary will be emphasised. Tables, graphs, and text will all be drawn up with these main results in mind. Other, less important, results will also be included, but their position and mass should indicate their relative lack of importance.

- Each *Discussion* section will also be constructed from the working summary using the principles we have already covered on page 112 for developing discussion topics.

The main purpose of the *Review of the Literature* is to provide background for, and to introduce, the hypotheses. The working summary is useful here, too, as a form of check list that can be used as a framework for the *Review of the Literature*.

... good writing and good science go hand in hand.

It would be surprising if, during the writing of the thesis, new ideas did not emerge. Such ideas can be incorporated into the working summary without reducing its effectiveness as an outline for the complete thesis. On the contrary, the working summary will assist you to weave new ideas into the fabric of the thesis by suggesting exactly where they should be included.

If we consider the working summary as the first draft of the thesis, then the expanded version we have just developed can be considered as the second draft. At this stage, a student writing a thesis should take advantage of having a supervisor who is, or should be, an officially appointed and readily available participant for the 'colleague test' (page 69). The experience of supervisors in writing papers and in supervision of other students, will be invaluable. Nevertheless, be careful that their familiarity with the work does not result in their missing badly phrased expressions and jargon. If you can obtain it, a second opinion, even on selected sections of the thesis, may be very helpful as a guide to the readability of your work. Make sure that colleagues whom you have induced to read sections of the thesis have access to your working summary so that they know where they are.

Many students find the writing of a thesis tedious and consider it an inordinate waste of time. In case you are tempted to think similarly, remember the principle expressed repeatedly throughout this book: good writing and good science go hand in hand. The training you are undertaking when writing your thesis is every bit as important as your research work. Treat it as such and your skill as a scientist will be enhanced. Your immediate colleagues may recognise you as a fine scientist and a thoroughly nice person through personal contact, but your standing with the other 99.99% of the scientific world will depend upon how well you write.

Index

A

abstract	49
acceptance of manuscripts	73
acknowledgements	87
ad libbing in talks	83
aim of an experiment	22, 24
anatomy of a thesis	114
appendices	62, 112
arguments	8
in Discussion	34, 39
article for non-scientists	102
attention of the audience	79
attitude	3
audience for posters	88
authority	46
authorship	51, 53

B

background in Introductions	27
bad posters	92
Bibliography	
in a thesis	113
brackets	62
breadth of a review	108
brevity	104
broad picture	107

C

catching the eye	
posters	89
category	
of arguments in Discussion	43
of results	33
check list for editing	68
check list for readability	70
choosing a journal	71
citations	48
clarity	4, 104
closing statement	87
clusters of nouns	56

co-authors	69
colleague test	69
'collection-of-papers' thesis	107
columns and rows	
in tables	37
complex adjectival phrases	57
components of an article	
for non-scientists	105
conclusion	40, 44
in a thesis	112
in discussion	40
in paragraph	40
in reviews	97, 98
in Title	18
Conclusions as sub-heading	41
consistency	49
content of posters	90
Contents	17
Contents page	17
context in Introductions	27
conversational style	
in oral presentations	80
in scientific writing	11
coping	
with editors	72
with referees	72
with reviewers	72
covering letter	72, 73
credibility	104
culling	
results	33

D

data in posters	91
dead time	84
defensible evidence	7
designing posters	89
difficulties with reviews	99
Discussion	8, 28, 39, 95
duplication	
in Results	38

SCIENTIFIC WRITING = THINKING IN WORDS

E

editing	10
editing for style	69
editors	5, 72, 73
editor's covering letter	74
Ehrenberg	36
ending your talk	87
English	
as a second language	11
conversational	11
evidence	46
expectation	23, 24, 64
in Title	18
experimental design	29
experimental procedure	111
expressions of confidence	46
eye contact	84

F

familiarity	69, 117
figures	34
final editing	68
flashy technology	79
fluency	55
focus	22, 24
follow-up	105, 106
footnotes	36
format of posters	90
funding bodies	53, 100

G

general hypothesis	116
general introduction	109
generalisations	46
in reviews	99
good posters	92
Gopen and Swan	64
graphic designers	90
graphs or tables?	36

H

headings	
row and column	35
headings in posters	91
headline	105
house rules	72
house style	11, 71
humour in oral presentations	81
hypothesis	7, 21, 101
accepting	8
formulation	24
in reviews	97
justification	21
latent	22
rejecting	9
specific	109, 112
unifying	109, 112

I

ideas	47, 95, 96
impact factors	71
impact in discussion	42
imprecise words	60
improvising in talks	83
ingredients for a	
popular article	104
instructions to authors	54
intellectual base	21
Introduction	7, 8, 20
in a review	96
scope	27

J

jargon	56
journal	
choosing	11, 71
impact factor	71
justification in posters	89
justifying arguments	41

K

key message in an oral presentation   79, 84

key words   18, 19

L

laboured humour   81

language of science   11

length of an article   103

length of the Discussion   47

linking words   64, 65

literature
   in a review   98

logical development   44, 45

logic in the Discussion   49

loose ends   44

M

major modifications   74

matching writing with reading   55

Materials and Methods   28
   'skimming'   28

methodology   21, 23, 30, 101

minor modifications   74

missing data   36

modern review   95

multi-authored papers   51

N

native English speakers   5, 55

'natural' scientific order   102

non-native English speakers   12, 55

non-scientists   100, 101

noun clusters   57

numerical data
   in tables   36

O

objective of an experiment   24

objectives of posters   89

objectivity of Results   31

opening of an oral presentation   79

opening sentence   79

opening statement   80

order of authorship   52

overfamiliarity   68

P

pacing a presentation   83

paragraph   40, 44
   in a Review   98

patterns
   in tables   35, 36, 37

peer-review   73

peer-reviewed journal   46

performance   78

personal communication   95

planning   6

position of arguments in Discussion   43

posters   88

précis   49

precision   4, 35, 104
   in tables   35

prediction   24

prepositions   56

priority
   of arguments   42
   of results   33, 111

proceedings   85

Q

qualifying clauses   58

questionnaire   21, 23

R

readability   55

reader expectation   46, 64, 105

readers   4

reading
   in oral presentations   83
   text verbatim   84

reasoning   24, 25

recommendation   40

referees 72

references 48, 53
  in a thesis 113

rehearsing a talk 84

rejection 71

rejection of manuscripts 74

relative importance
  of results 33

repeating 'old' information 64

repetition of data 47

replacing nouns with verbs 59

research grant 53

re-submitting to the journal 74

Results 8, 31, 101
  in Title 18
  separating from Discussion 31

Results and Discussion 31

Review 95

reviewers 5

review of the literature
  in a thesis 107

S

scientific interest in posters 89

scientific method 7

scientific proposal 21

search programs
  electronic 17

sending to the journal 72

shortening sentences 59

signpost words 65

size of font for posters 91, 92

size of paragraphs 41, 43

size of Summary 49

specific hypotheses 109

speculation 40, 47, 97

standard deviation 39

standard error 39

statistical analysis
  in Methodology 30

statistics 2, 38

sticking to time 86

structure 8, 12, 16

of an oral presentation 78

of a review 96

of posters 90

physical 16

scientific 13

style 4, 5, 55
  conversational 5

sub-headings
  in Discussion 45
  in Materials and Methods 28

Subject matter
  in articles for the public 103

subordinate clauses 58

Summary 49, 105
  components 50
  constructing 50
  in a thesis 113
  in oral presentation 82

supervisor 116, 117

survey 21

T

tables 34
  for precision 35

tables or graphs? 36

take-home message 80, 86, 105

talking about posters 92

text
  for clarity 35
  in Results 34

timing in articles for the public 103

timing of posters 88

Title 17, 105
  long 18

topic sentence 44

typographical errors 72

U

'unified' thesis 107

unifying arguments 108

unifying hypothesis 109, 112

unimaginative posters 90

unpublished data 95
unreliable information 99

V

validation 111
validation of new techniques 30
variety of style 82
verbal stumbling blocks 55, 56

W

working summary
  for a thesis 115

Y

'you' in oral presentations 81
young authors 71